Telling Time

Clocks Workbook

Hours · Half Hours · Quarter Hours · 5-Minute Marks · 1-Minute Marks

by
Stacy Otillio & Frank Otillio

TABLE OF CONTENTS

Word Problems are included at the end of each section.

SECTION 1

Times Ending in **:00 and :30**

Worksheet Types

Match Clocks to Times
Read the Time, Write the Time, Draw the Hands
Future Time, Past Time
Word Problems

Name ______________________ Date ______________

Times Ending in :00 and :30

Draw lines to match the clocks to the correct times.

11:00

4:00

7:30

12:30

6:30

Name ______________________ Date ______________________

Times Ending in :00 and :30

Write the time in the blanks below the clocks.

Draw the hands on the clocks to match the times.

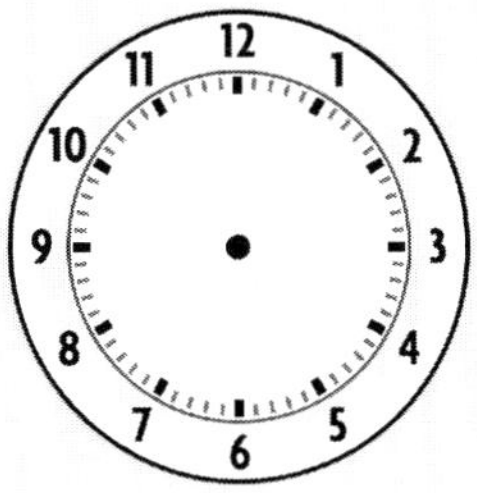   

11:30	6:00	2:30	9:00

Name ______________________ Date ______________

Times Ending in :00 and :30

Write the time in the blanks below the clocks.

Draw the hands on the clocks to match the times.

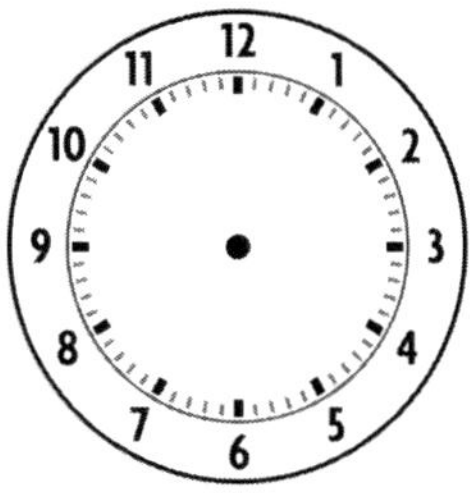
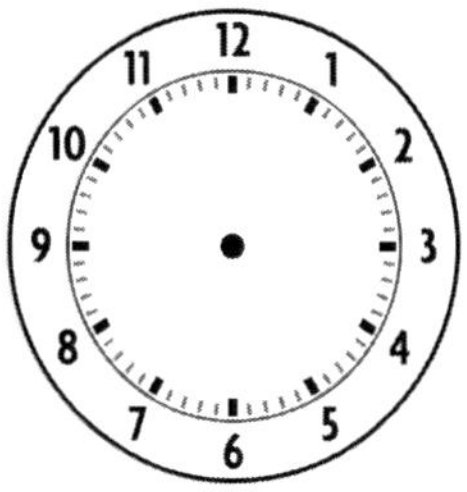
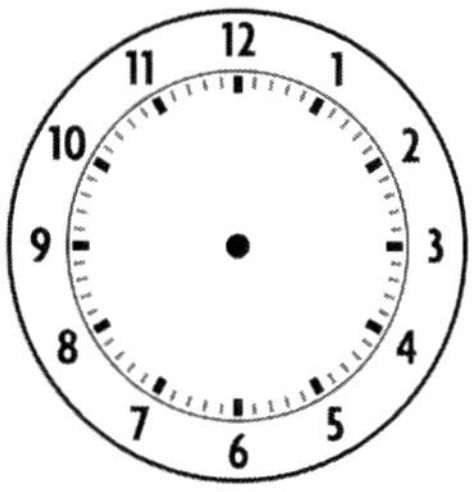
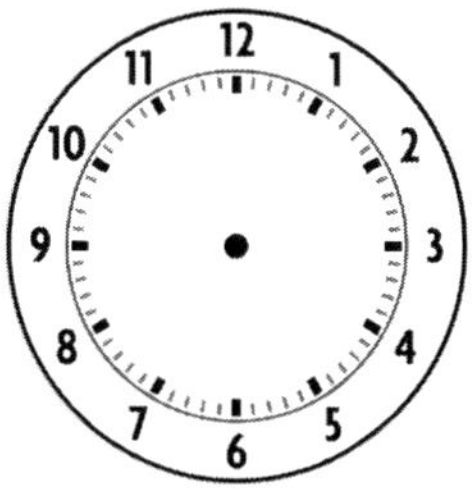

5:00	6:30	4:00	1:30

Name ________________________________ Date ____________________

Times Ending in :00 and :30

Write the time in the blanks below the clocks.

Draw the hands on the clocks to match the times.

5:30 1:00 10:00 3:30

Name ______________________ Date ______________

Times Ending in :00 and :30

Write the time below the clock showing the hands, then write the correct future time below the empty clock and draw the hands.

30 minutes later

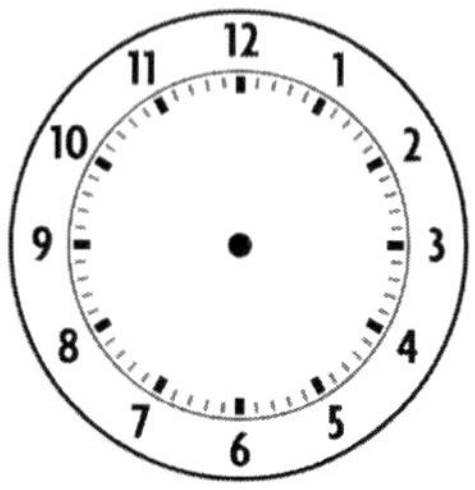

1 hour later

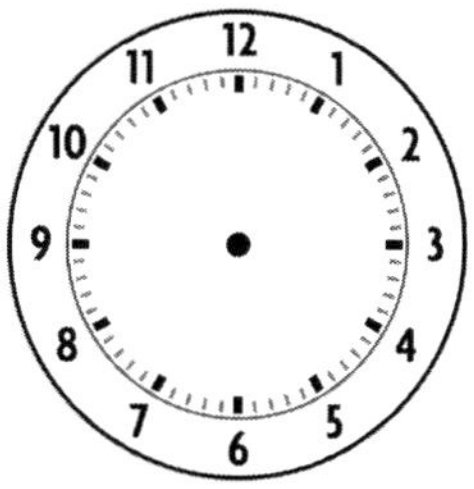

Write the time below the clock showing the hands, then write the correct past time below the empty clock and draw the hands.

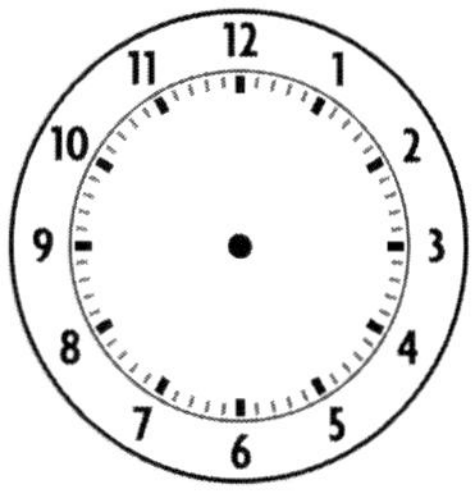

1 hour earlier

Name ____________________ Date ____________________

Times Ending in :00 and :30

Write the time below the clock showing the hands, then write the correct future time below the empty clock and draw the hands.

2 hours later

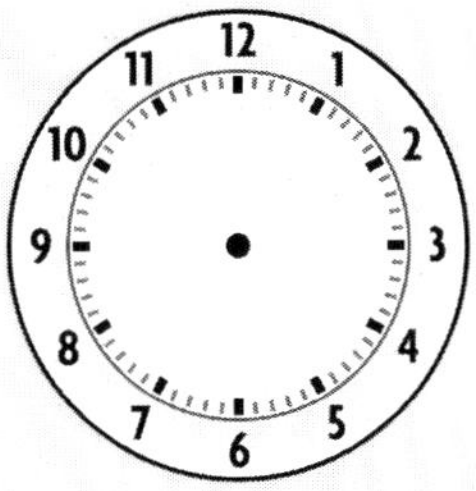

30 minutes later

Write the time below the clock showing the hands, then write the correct past time below the empty clock and draw the hands.

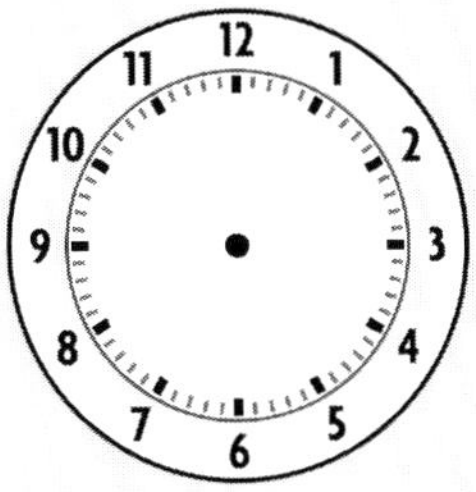

30 minutes earlier

Name ______________________________ Date ____________________

Times Ending in :00 and :30

For each event, write the beginning and ending times under the clocks, then draw the hands on the clocks to match the correct times.

1. Emma wakes up at 6:00 and arrives at school 2 hours later.

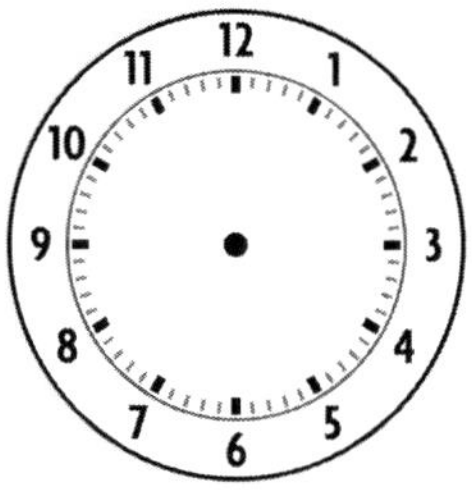
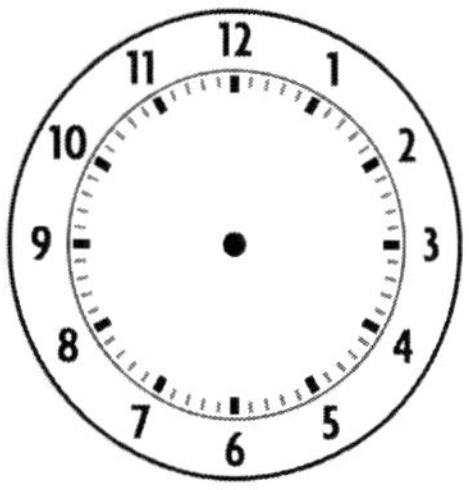

2. Greg's soccer practice starts at 4:00 and lasts 1 hour.

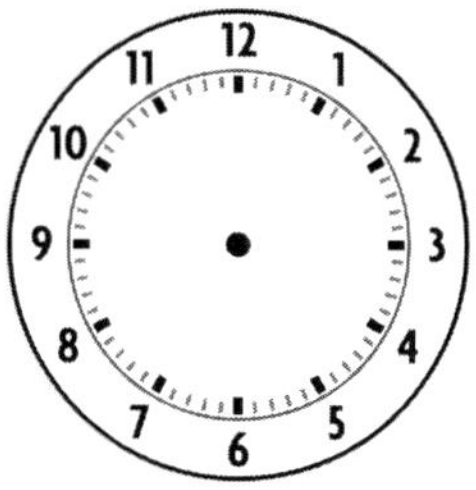
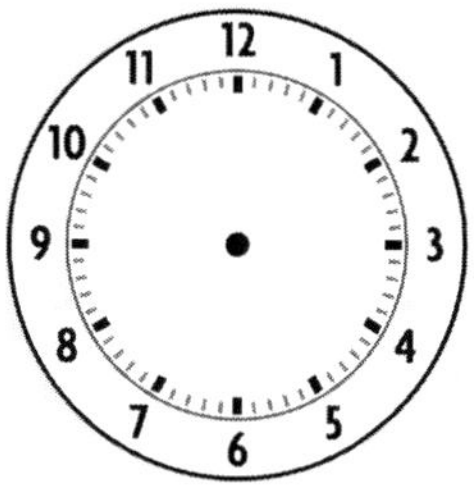

3. Lunch begins at 12:00 and lasts 30 minutes.

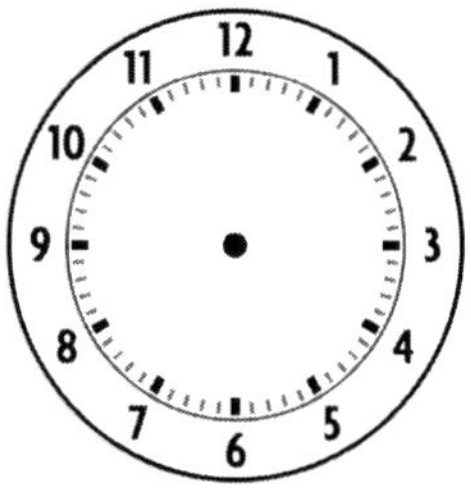
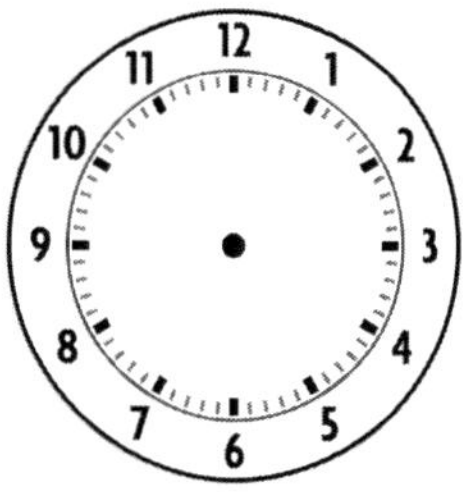

4. James played video games for 1 hour and finished playing at 10:00.

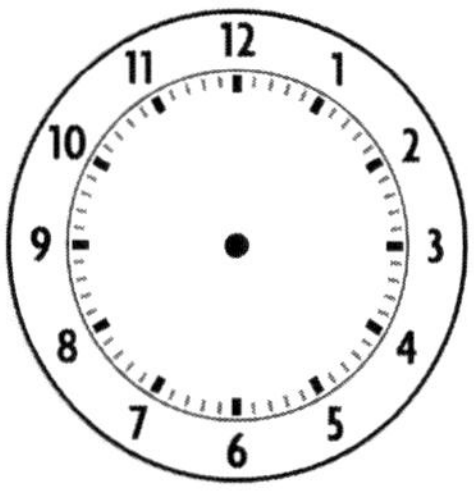
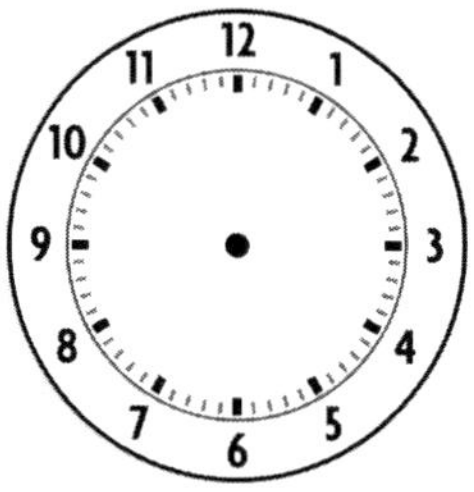

SECTION 2

Times Ending in :15

Worksheet Types

Read the Time, Write the Time, Draw the Hands
Future Time, Past Time
Word Problems

Name ______________________ Date ______________

Times Ending in :15

Write the time in the blanks below the clocks.

Draw the hands on the clocks to match the times.

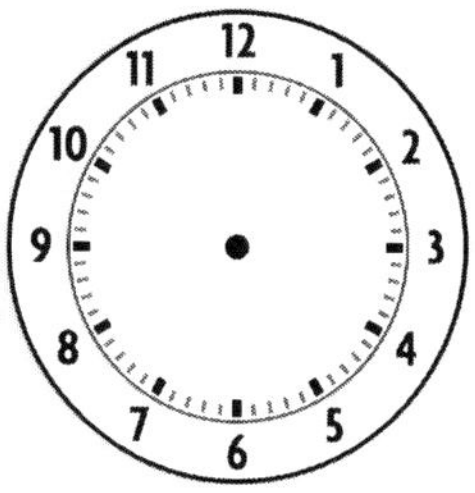
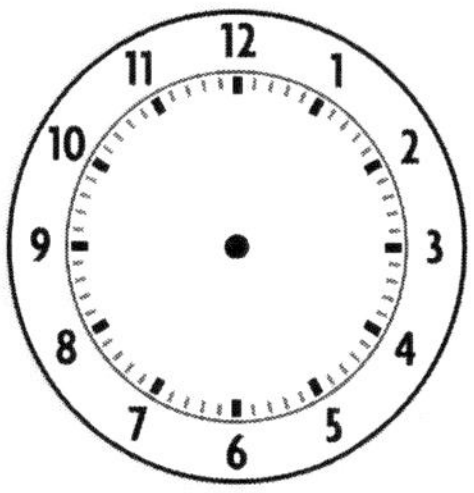

4:15	10:15	6:15	2:15

Name ______________________ Date ______________

Times Ending in :15

Write the time below the clock showing the hands, then write the correct future time below the empty clock and draw the hands.

15 minutes later

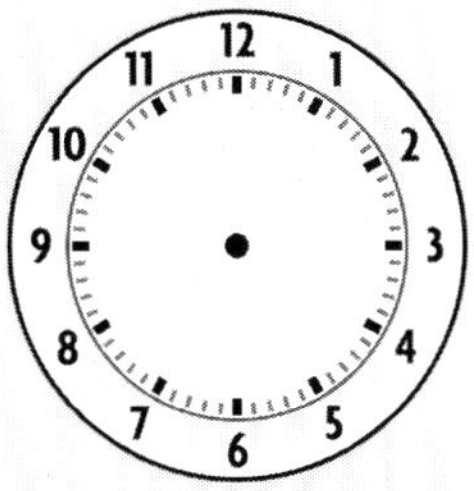

15 minutes later

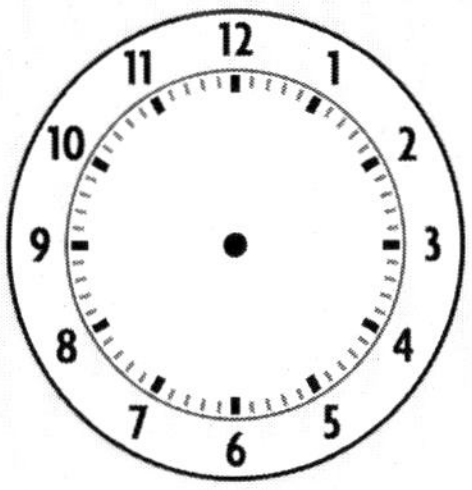

Write the time below the clock showing the hands, then write the correct past time below the empty clock and draw the hands.

15 minutes earlier

Name ______________________ Date ______________

Times Ending in :15

For each event, write the beginning and ending times under the clocks, then draw the hands on the clocks to match the correct times.

1. John leaves at 7:00 and gets to school 15 minutes later.

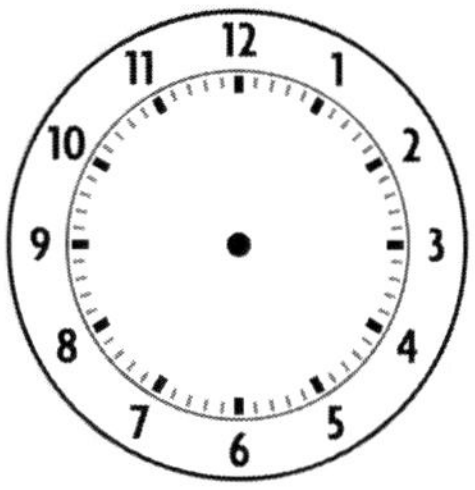
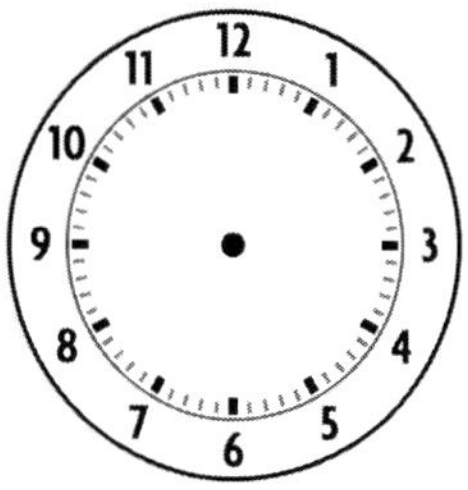

2. It is 3:00 and Julie's art class begins in 15 minutes.

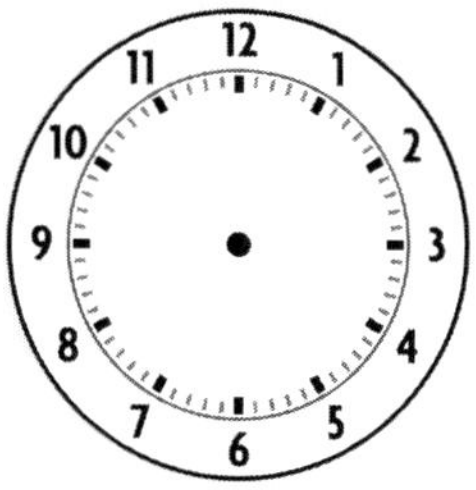
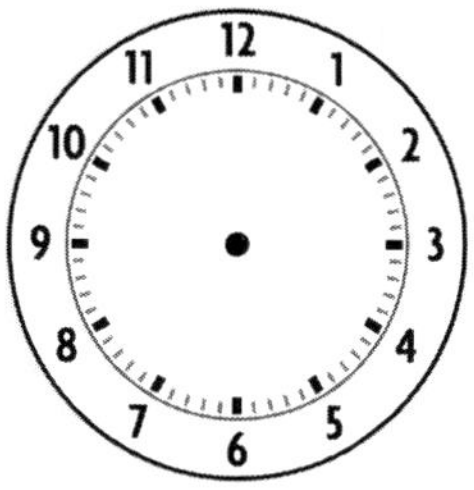

3. Basketball practice begins at 4:15 and lasts 1 hour.

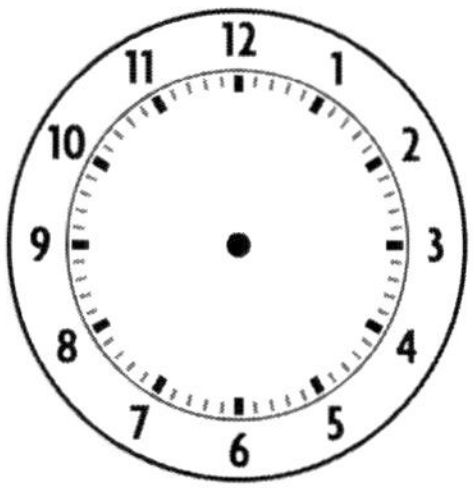
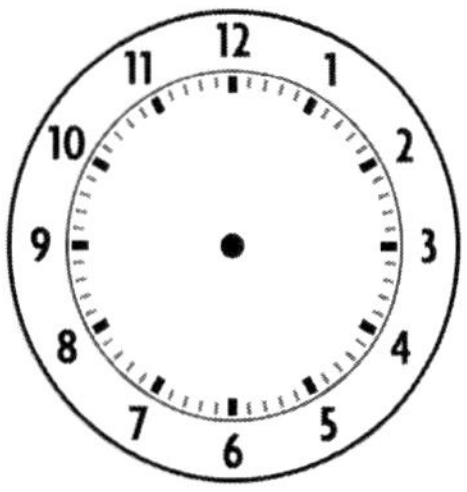

4. Olivia jogs for 15 minutes. She finishes at 10:15.

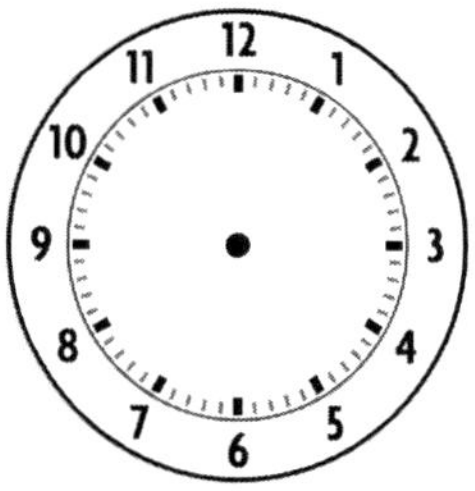
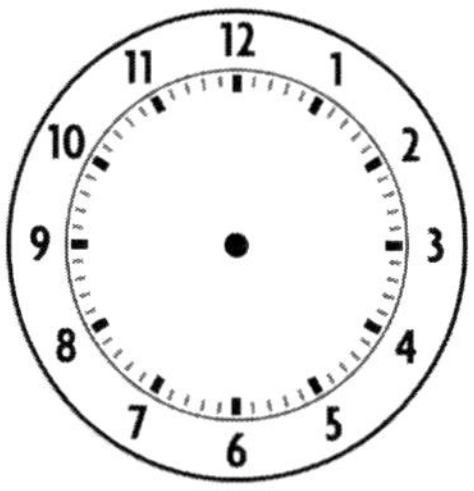

SECTION 3

Times Ending in :00, :15, :30

Worksheet Types

Match Clocks to Times
Read the Time, Write the Time, Draw the Hands
Future Time, Past Time
Word Problems

Name ______________________ Date ______________

Times Ending in :00, :15, :30

Draw lines to match the clocks to the correct times.

1:00

4:15

3:30

11:00

7:15

Name ______________________ Date ______________

Times Ending in :00, :15, :30

Draw lines to match the clocks to the correct times.

6:30

3:00

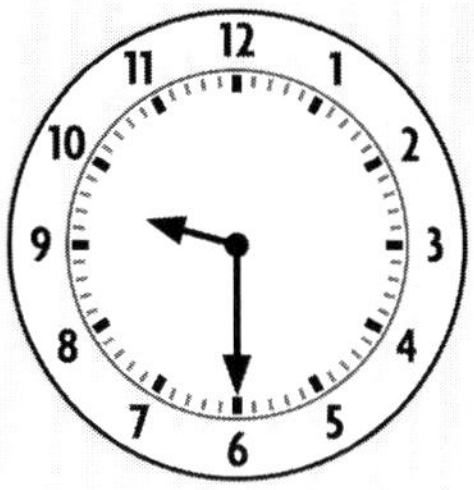

2:15

9:30

10:15

Name ____________________ Date ____________________

Times Ending in :00, :15, :30

Write the time in the blanks below the clocks.

Draw the hands on the clocks to match the times.

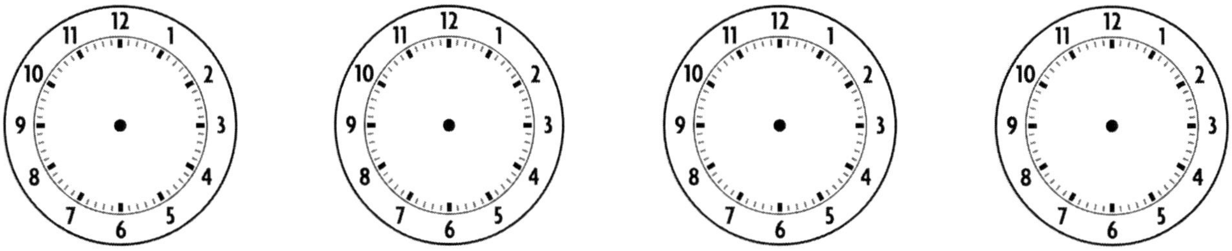

4:30	7:00	8:15	1:15

Name ______________________________ Date ____________________

Times Ending in :00, :15, :30

Write the time in the blanks below the clocks.

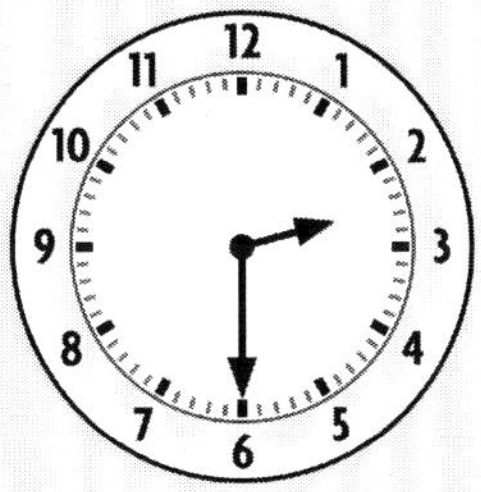

Draw the hands on the clocks to match the times.

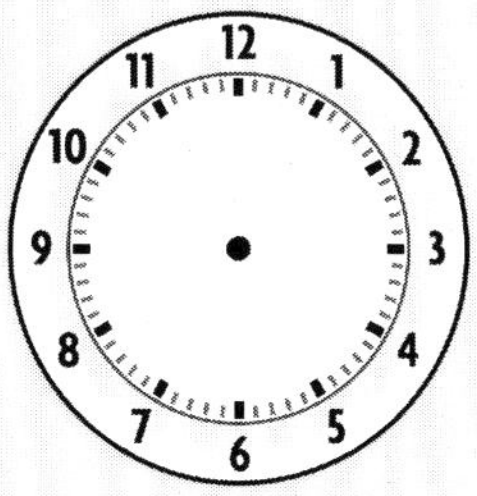
6:00

11:30

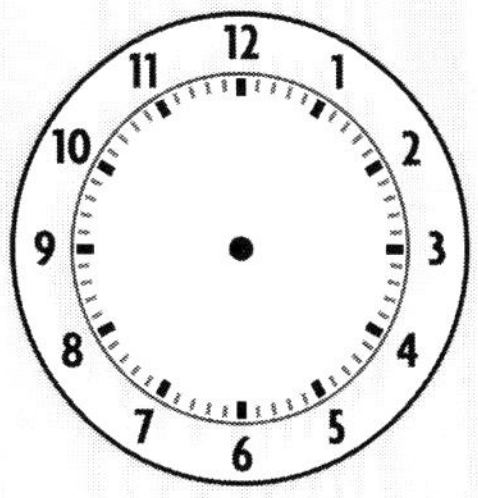
10:15

5:30

Name ______________________________ Date ____________________

Times Ending in :00, :15, :30

Write the time in the blanks below the clocks.

Draw the hands on the clocks to match the times.

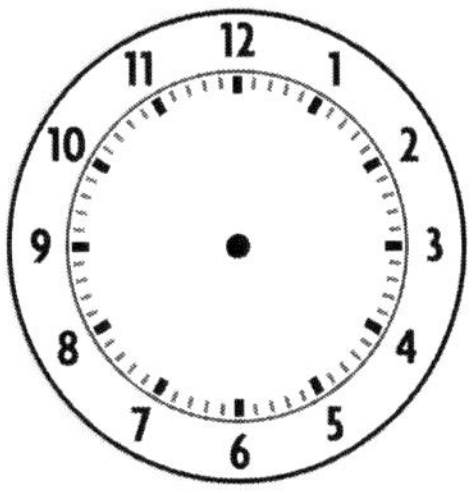  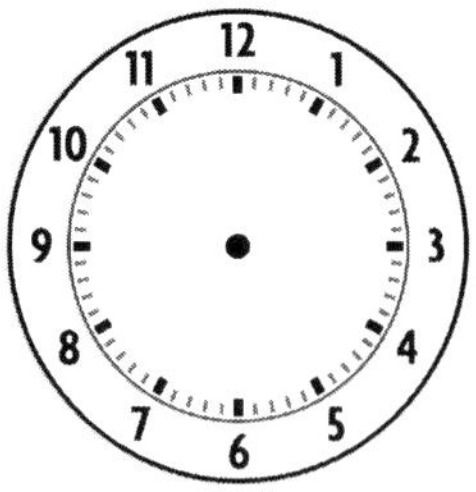 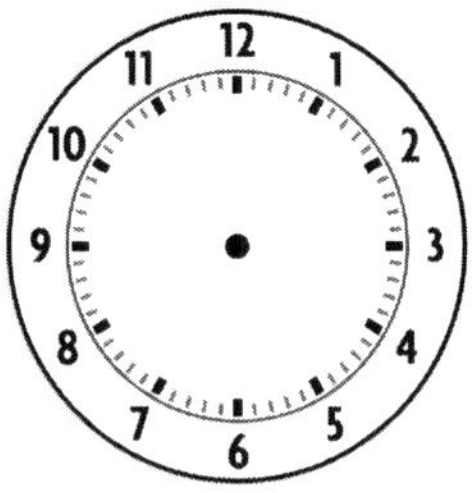

7:15 | 6:30 | 11:15 | 9:00

Name ______________________ Date ______________

Times Ending in :00, :15, :30

Write the time in the blanks below the clocks.

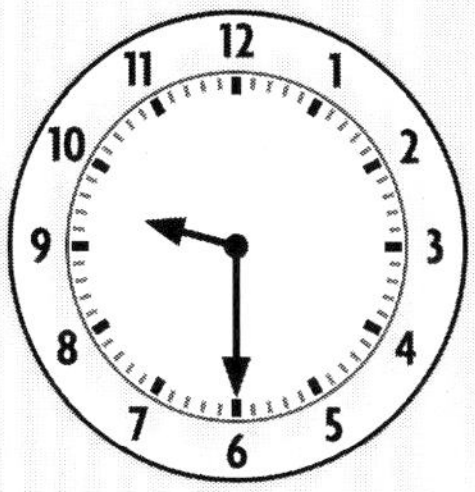

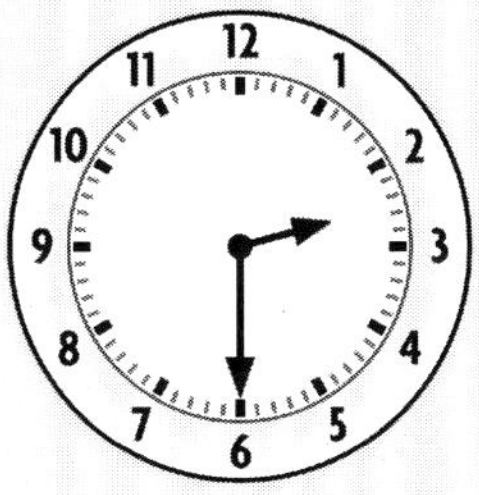

Draw the hands on the clocks to match the times.

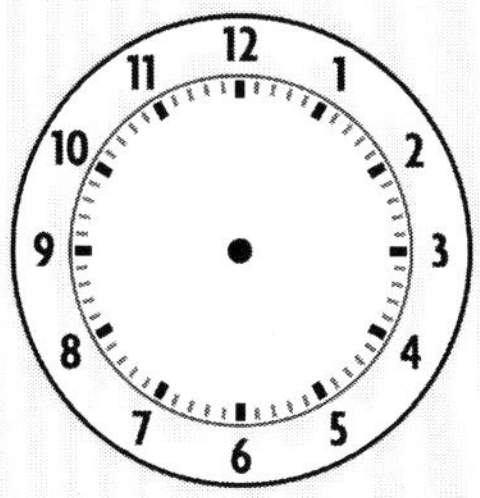

1:00

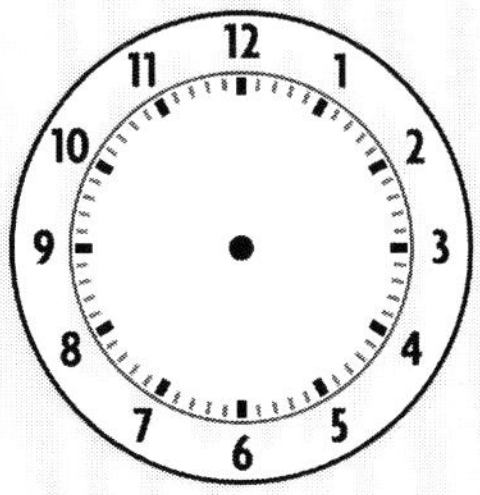

9:15

12:30

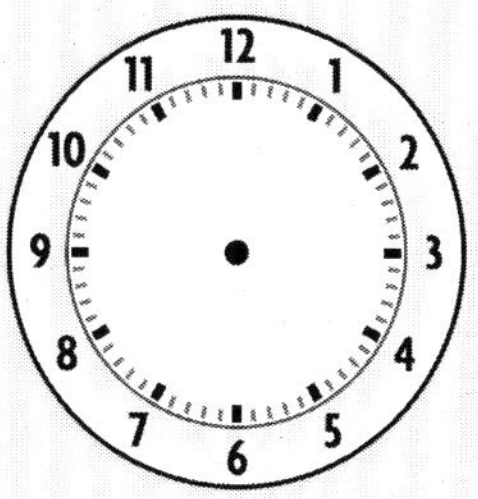

7:30

Name ______________________________ Date ____________________

Times Ending in :00, :15, :30

Write the time in the blanks below the clocks.

Draw the hands on the clocks to match the times.

3:15

2:30

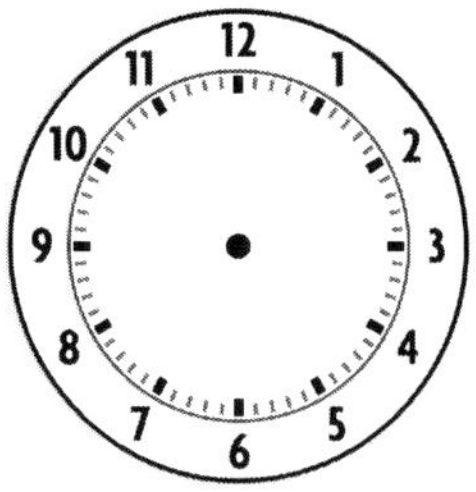
12:15

10:00

Name ______________________ Date ______________

Times Ending in :00, :15, :30

Write the time in the blanks below the clocks.

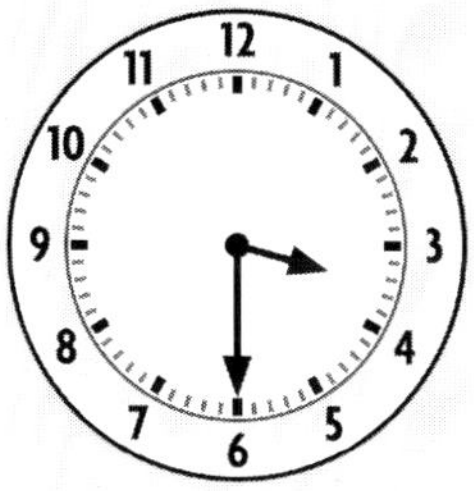

Draw the hands on the clocks to match the times.

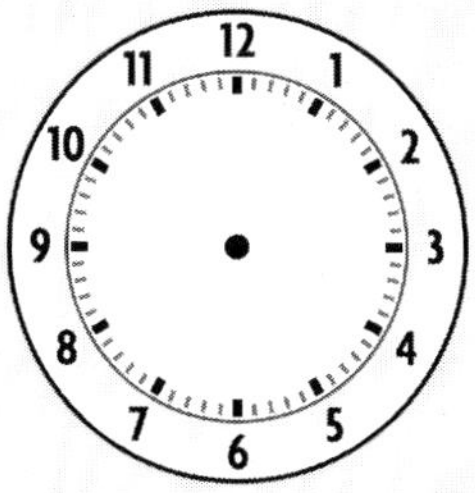
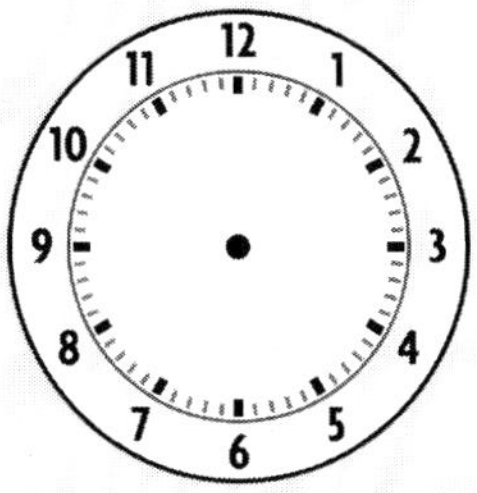
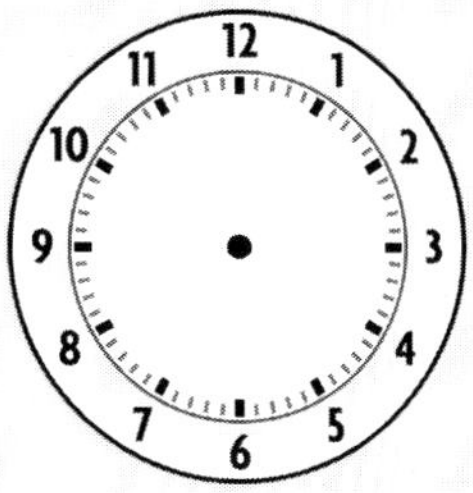

11:00 7:30 8:15 6:30

Name ________________________________ Date ____________________

Times Ending in :00, :15, :30

Write the time in the blanks below the clocks.

Draw the hands on the clocks to match the times.

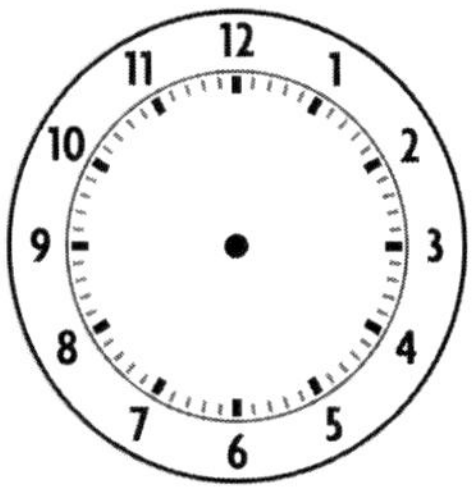 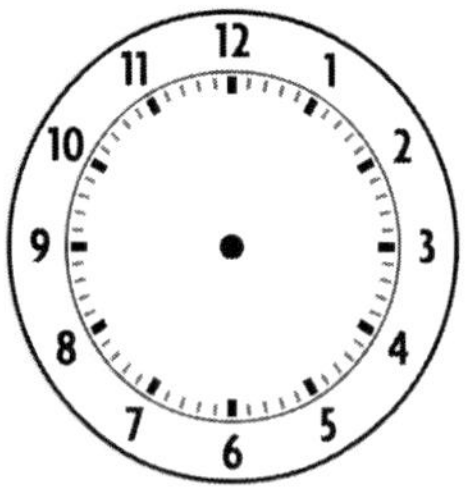 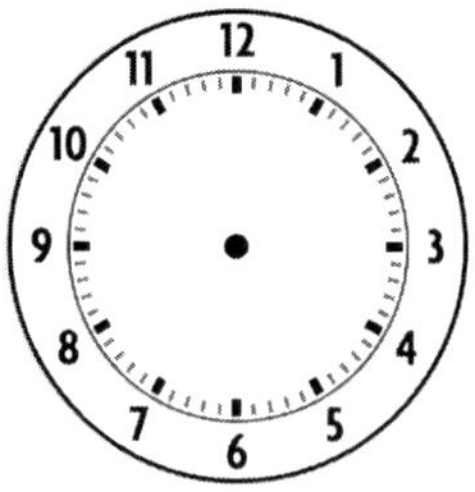 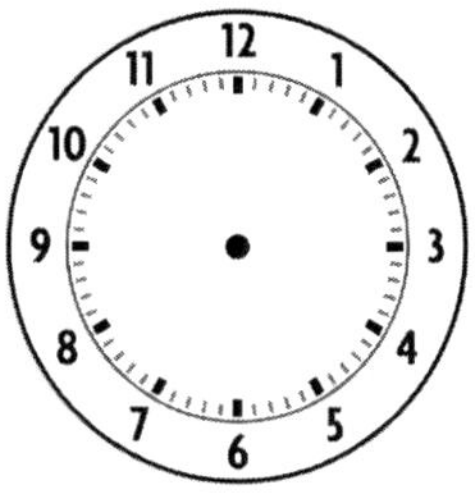

5:15	9:15	8:00	12:30

Name ______________________ Date ______________

Times Ending in :00, :15, :30

Write the time below the clock showing the hands, then write the correct future time below the empty clock and draw the hands.

30 minutes later

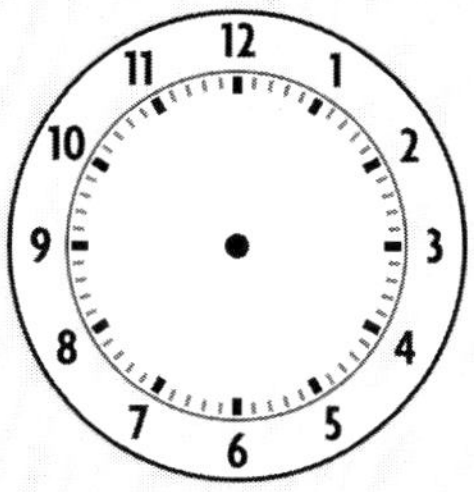

1 hour later

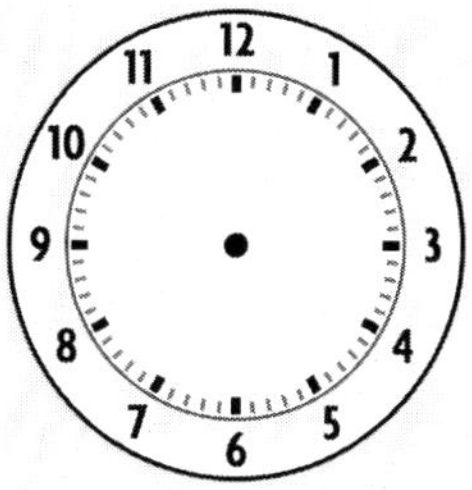

Write the time below the clock showing the hands, then write the correct past time below the empty clock and draw the hands.

30 minutes earlier

Name ______________________ Date ______________

Times Ending in :00, :15, :30

Write the time below the clock showing the hands, then write the correct future time below the empty clock and draw the hands.

2 hours later

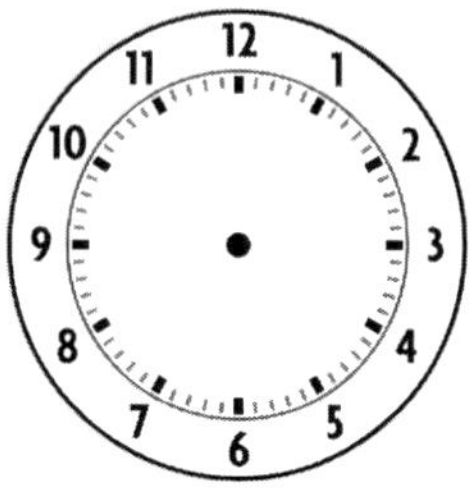

30 minutes later

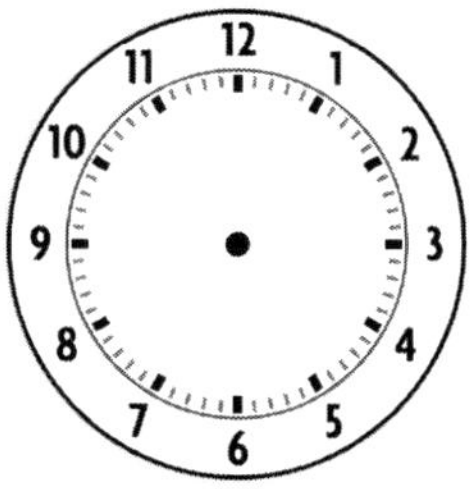

Write the time below the clock showing the hands, then write the correct past time below the empty clock and draw the hands.

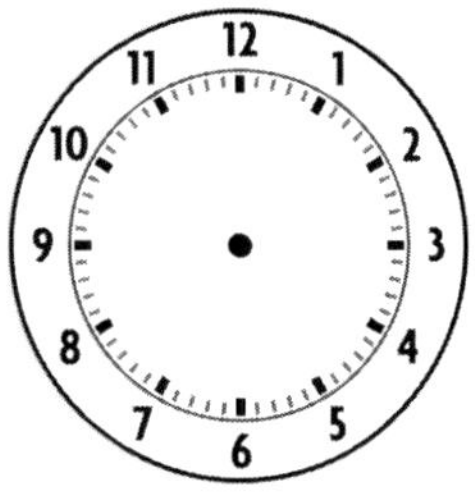

15 minutes earlier

Name ____________________ Date ____________________

Times Ending in :00, :15, :30

Write the time below the clock showing the hands, then write the correct future time below the empty clock and draw the hands.

1 hour later

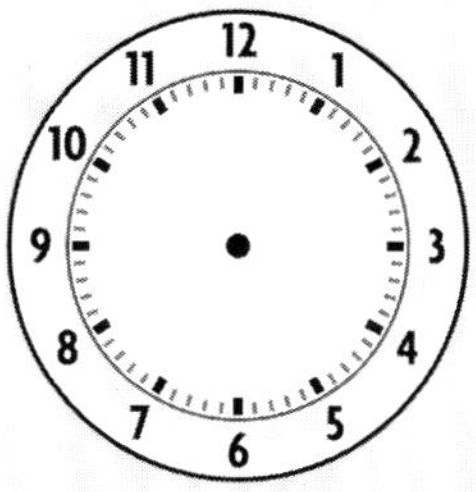

2 hours later

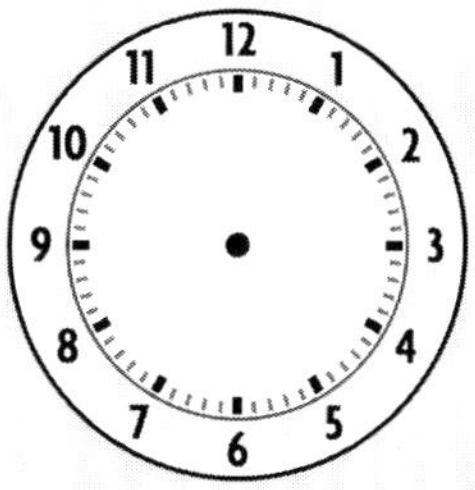

Write the time below the clock showing the hands, then write the correct past time below the empty clock and draw the hands.

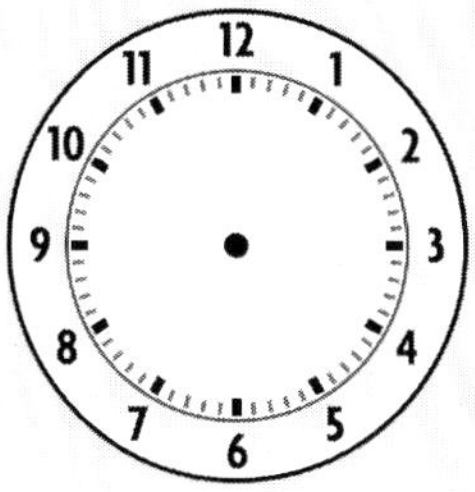

15 minutes earlier

Name ______________________________ Date ____________________

Times Ending in :00, :15, :30

For each event, write the beginning and ending times under the clocks, then draw the hands on the clocks to match the correct times.

1. The plane took off at 2:30 and landed 1 hour later.

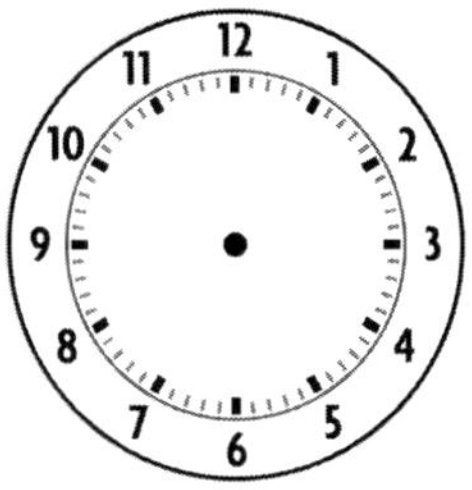

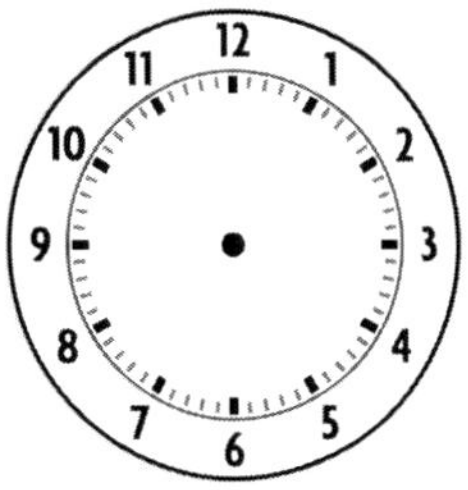

2. Alexis started eating at 3:00 and finished in 15 minutes.

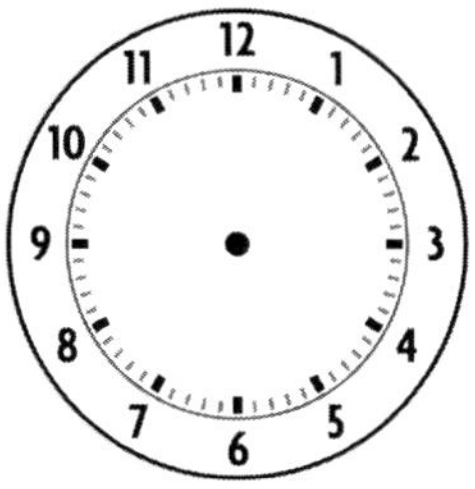

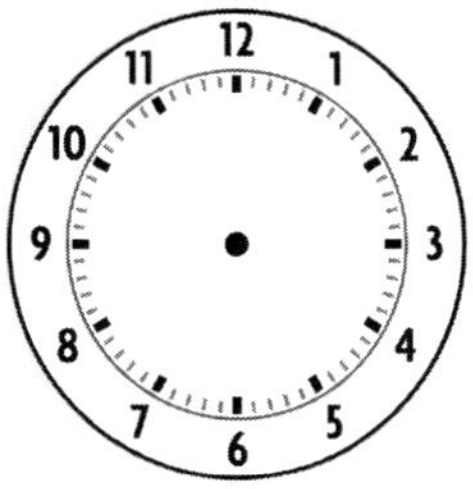

3. The party started at 4:15 and ended 2 hours later.

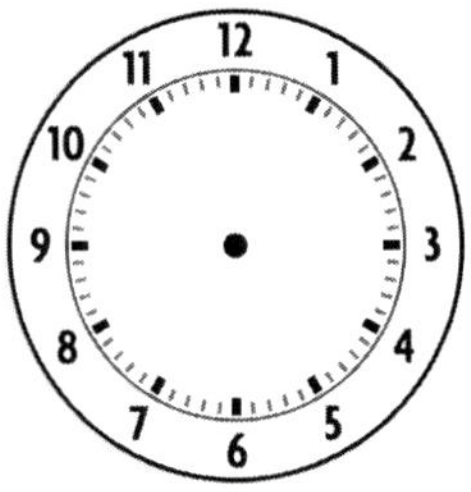

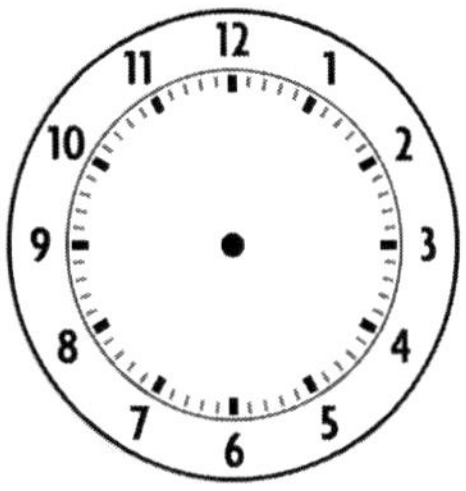

4. The children slept for 30 minutes and woke up at 5:30.

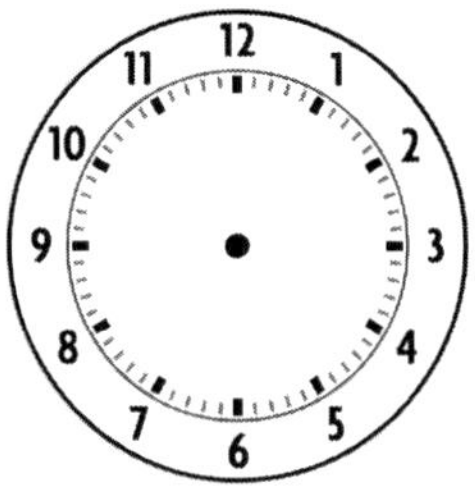

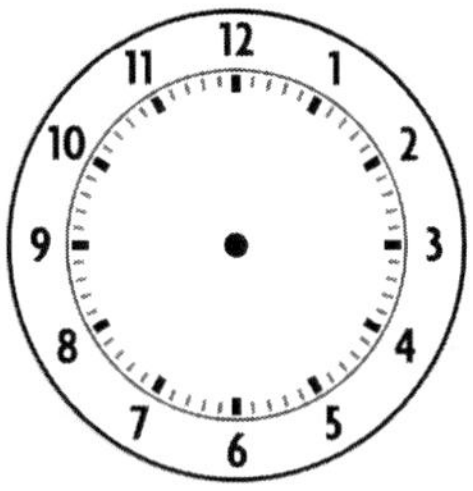

Name ______________________ Date ______________

Times Ending in :00, :15, :30

For each event, write the beginning and ending times under the clocks, then draw the hands on the clocks to match the correct times.

1. Ryan started his math homework at 4:00 and finished in 15 minutes.

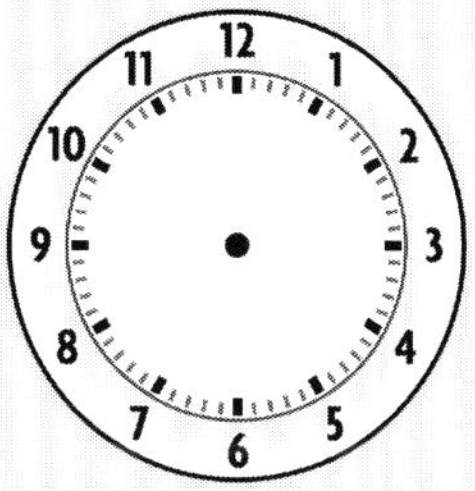

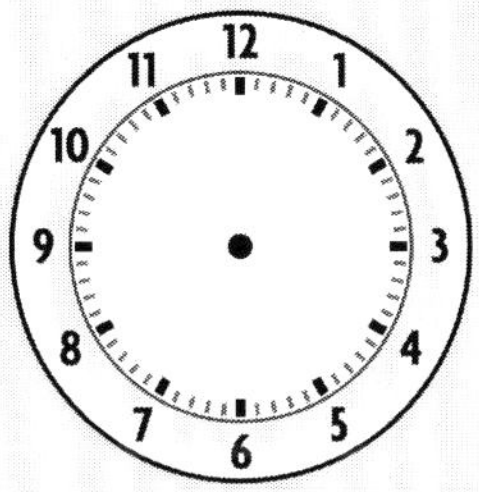

2. Melissa left her house at 2:00 and walked for 30 minutes.

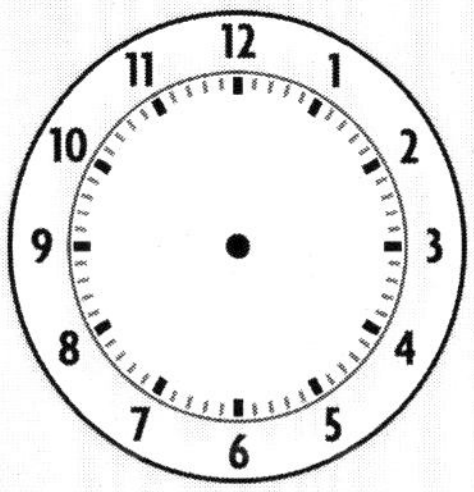

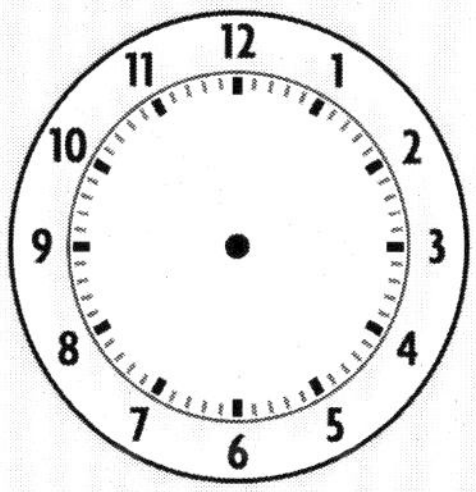

3. Jada's dance class begins at 6:15 and lasts 1 hour.

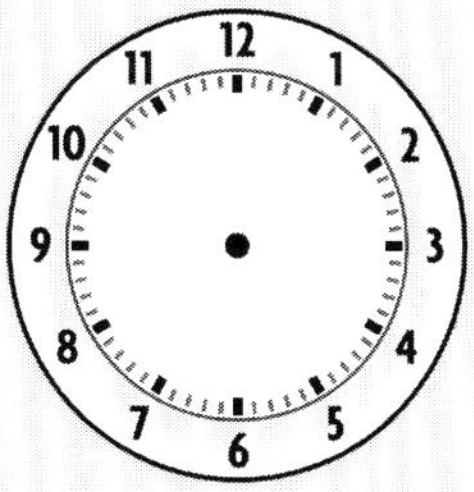

4. The band played for 30 minutes. They finished at 3:30.

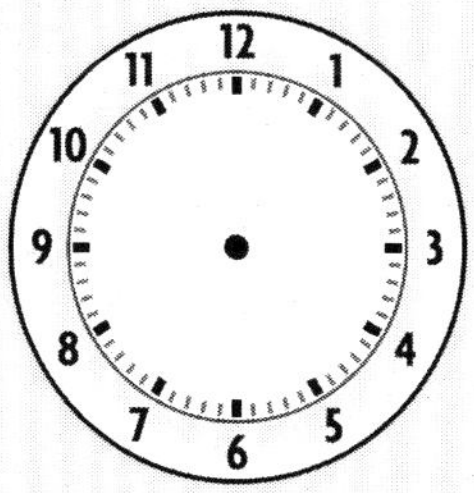

Name ______________________________ Date ____________________

Times Ending in :00, :15, :30

For each event, write the beginning and ending times under the clocks, then draw the hands on the clocks to match the correct times.

1. Brianna left work at 5:00 and arrived home 15 minutes later.

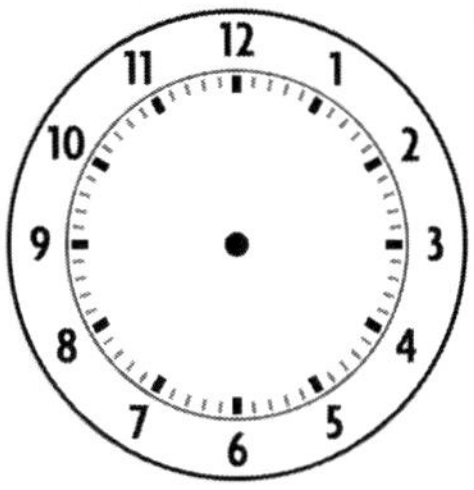
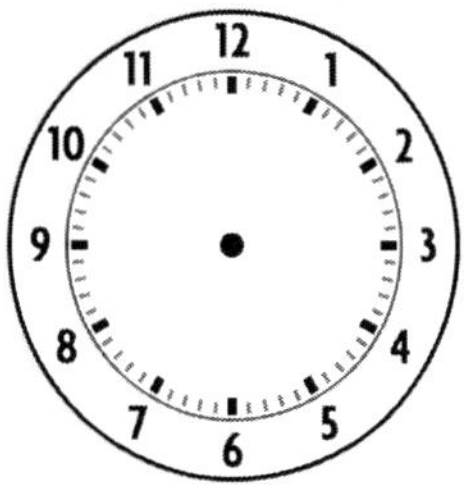

2. Angele began her homework at 2:30 and finished 1 hour later.

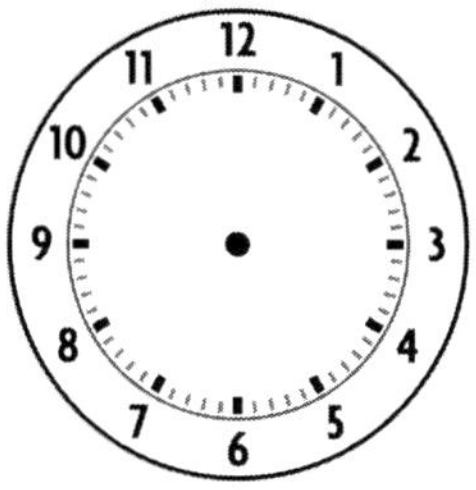
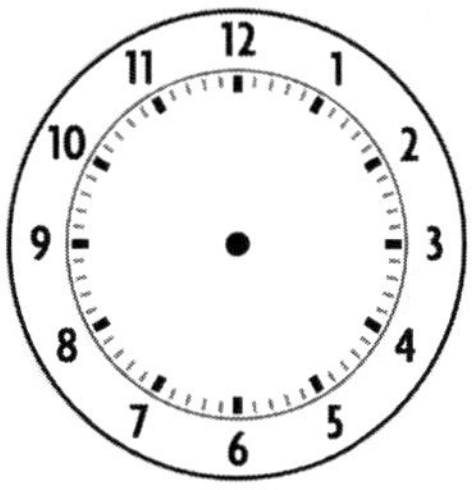

3. Luke started painting at 12:00 and finished in 15 minutes.

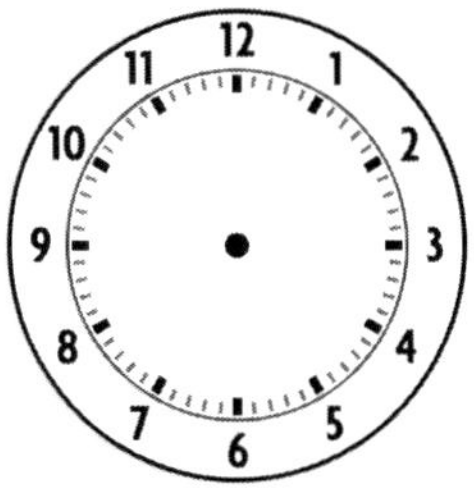
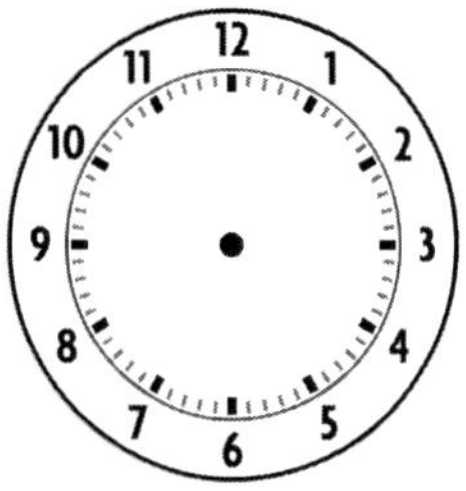

4. Ronny played a game for 15 minutes. He stopped playing at 6:15.

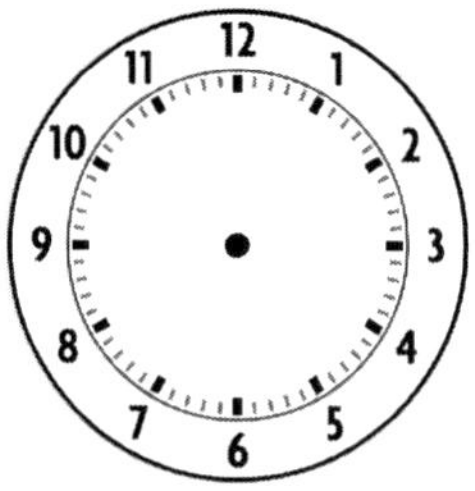
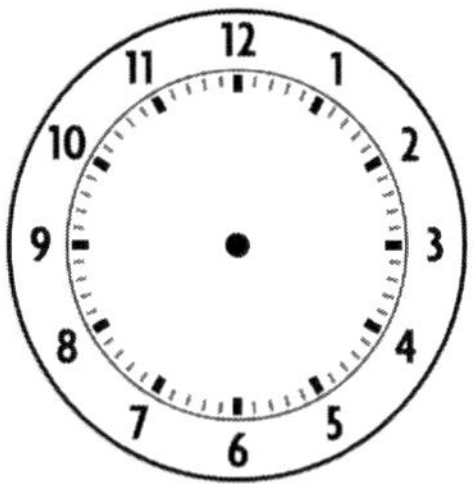

SECTION

4

Times Ending in :45

Worksheet Types

Read the Time, Write the Time, Draw the Hands
Future Time, Past Time
Word Problems

Name ______________________ Date ______________

Times Ending in :45

Write the time in the blanks below the clocks.

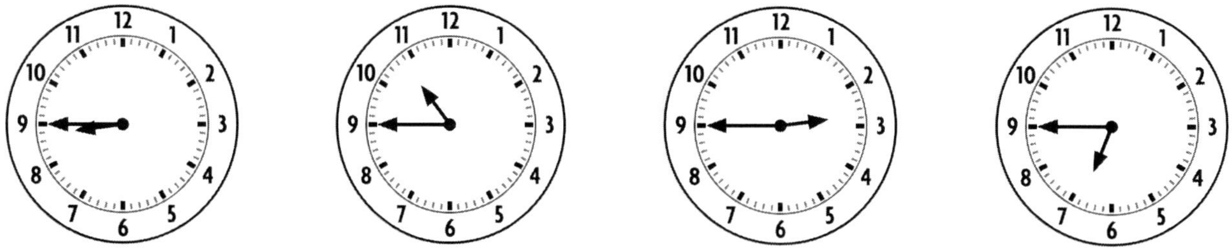

Draw the hands on the clocks to match the times.

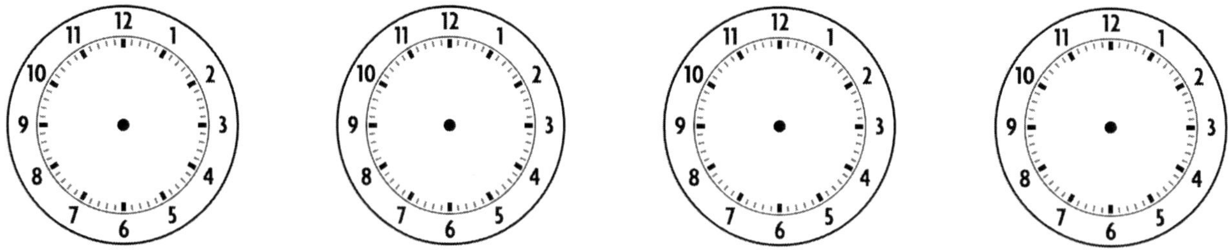

1:45	12:45	4:45	7:45

Name ______________________ Date ______________

Times Ending in :45

Write the time below the clock showing the hands, then write the correct future time below the empty clock and draw the hands.

45 minutes later

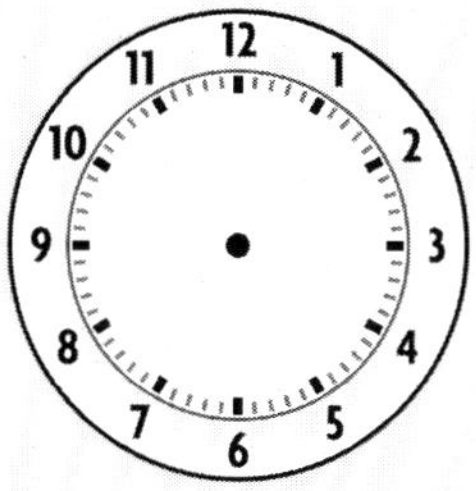

45 minutes later

Write the time below the clock showing the hands, then write the correct past time below the empty clock and draw the hands.

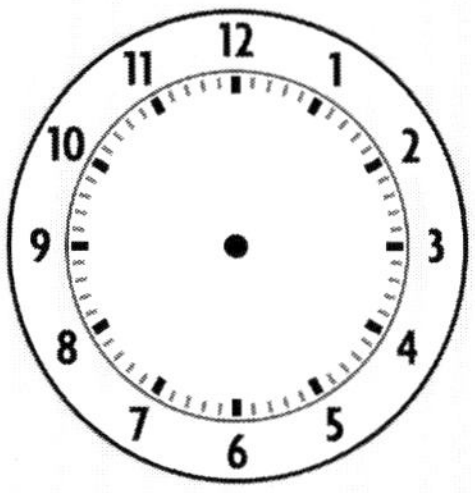

45 minutes earlier

Name ______________________________ Date ____________________

Times Ending in :45

For each event, write the beginning and ending times under the clocks, then draw the hands on the clocks to match the correct times.

1. Trey starts drawing at 5:00 and he draws for 45 minutes.

2. The bus left at 7:00 and stopped at Jason's house 45 minutes later.

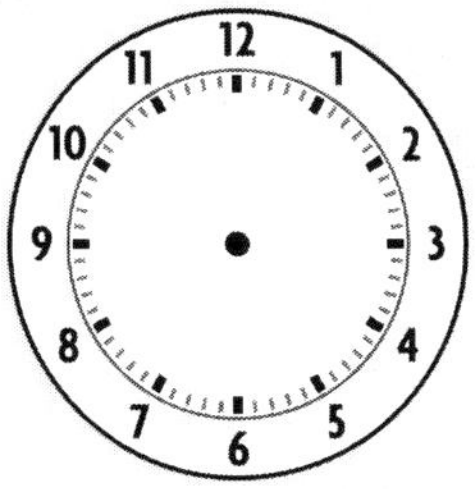

3. Lydia's cat took a nap at 1:45 and woke up 1 hour later.

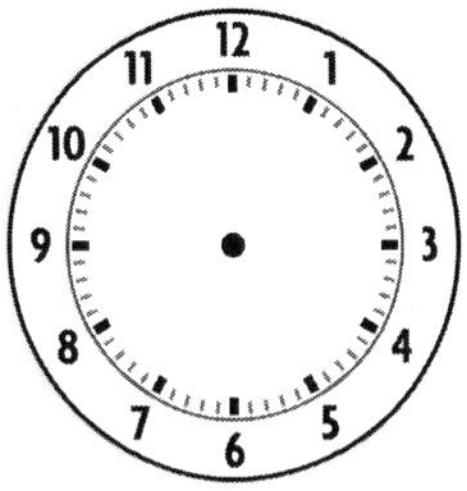

4. The play was 45 minutes long and ended at 6:45.

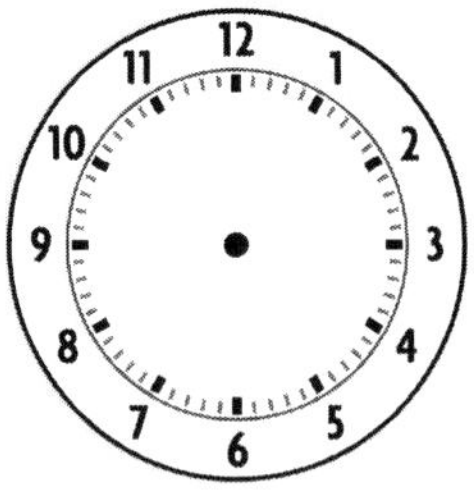

SECTION 5

Times Ending in :00, :15, :30, :45

Worksheet Types

Match Clocks to Times
Read the Time, Write the Time, Draw the Hands
Future Time, Past Time
Word Problems

Name ______________________ Date ______________

Times Ending in :00, :15, :30, :45

Draw lines to match the clocks to the correct times.

9:30

12:15

3:45

9:00

5:45

Name ______________________________ Date ____________________

Times Ending in :00, :15, :30, :45

Draw lines to match the clocks to the correct times.

10:45

7:15

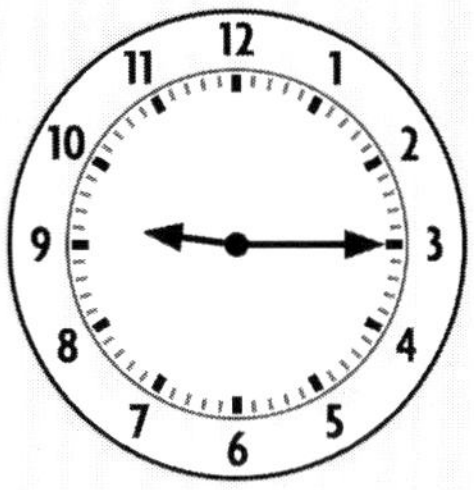

4:30

6:45

9:15

Name ______________________________ Date ____________________

Times Ending in :00, :15, :30, :45

Write the time in the blanks below the clocks.

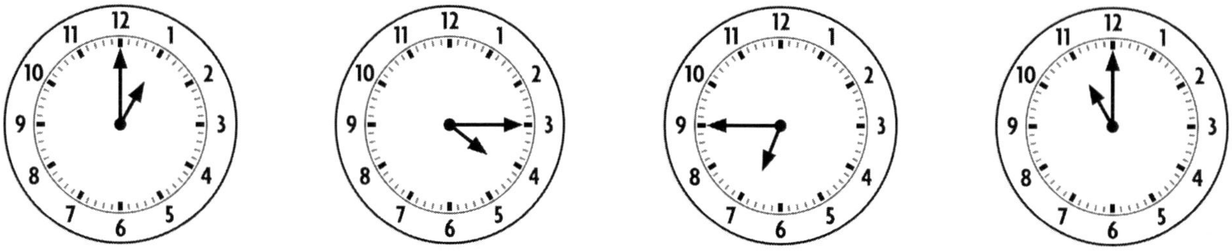

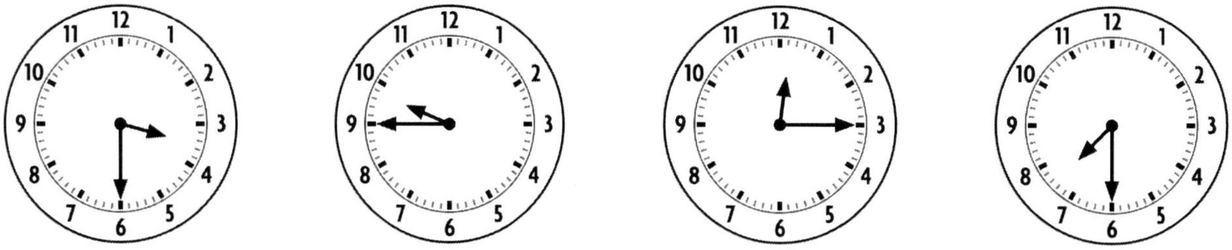

Draw the hands on the clocks to match the times.

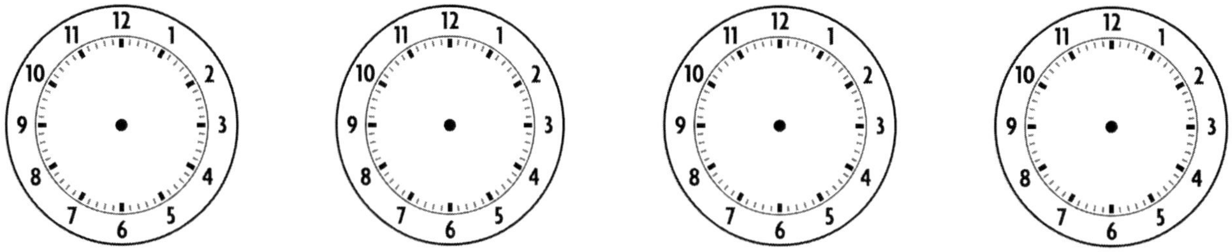

5:00	8:15	2:45	10:30

Name ________________________ Date ________________

Times Ending in :00, :15, :30, :45

Write the time in the blanks below the clocks.

Draw the hands on the clocks to match the times.

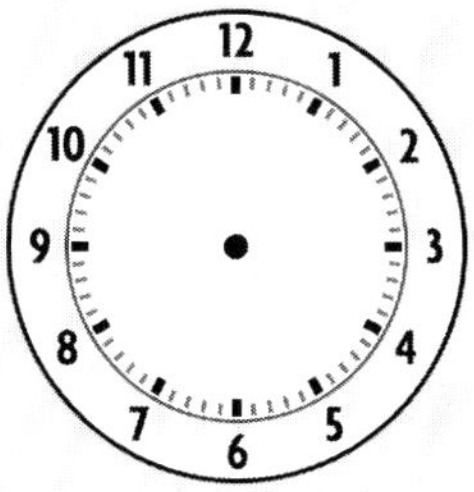

10:45

4:30

2:15

9:00

Name ______________________________ Date ____________________

Times Ending in :00, :15, :30, :45

Write the time in the blanks below the clocks.

Draw the hands on the clocks to match the times.

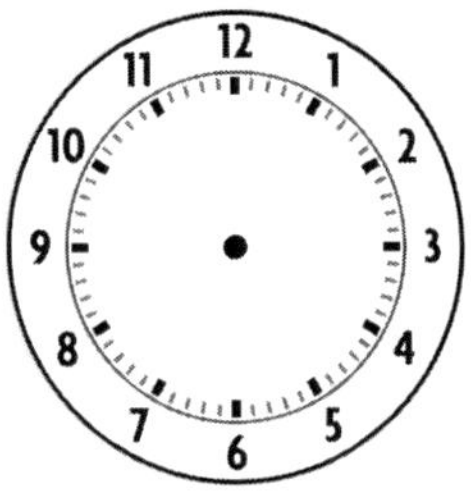 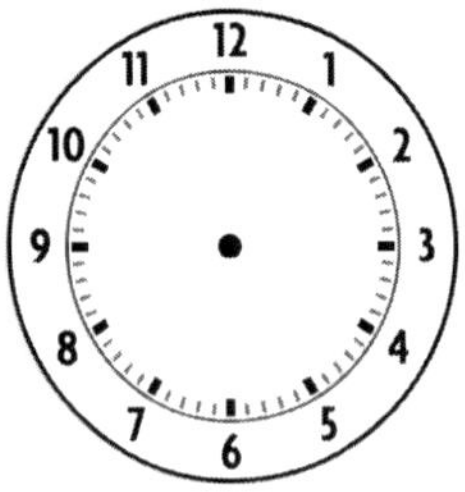 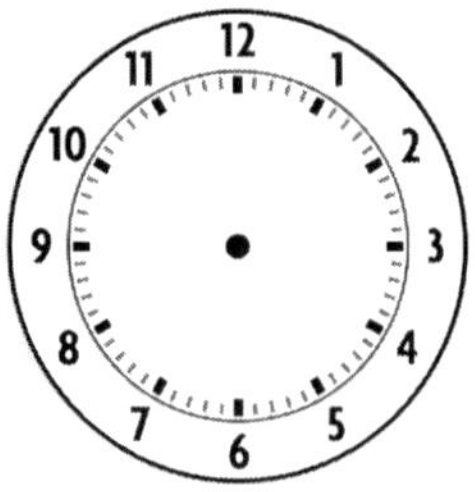 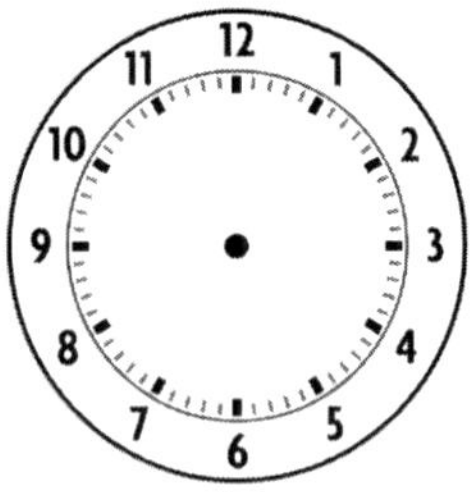

5:30	9:15	4:45	1:30

Name ______________________________ Date ________________

Times Ending in :00, :15, :30, :45

Write the time in the blanks below the clocks.

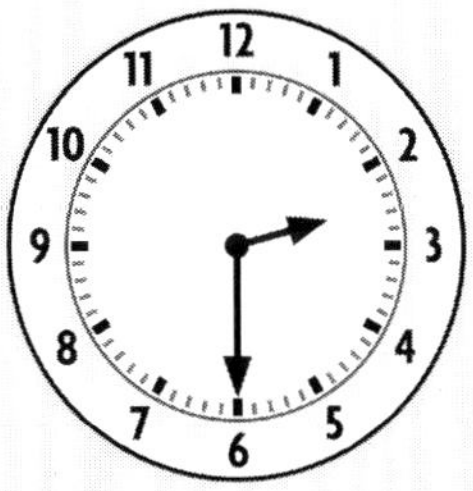

Draw the hands on the clocks to match the times.

1:15

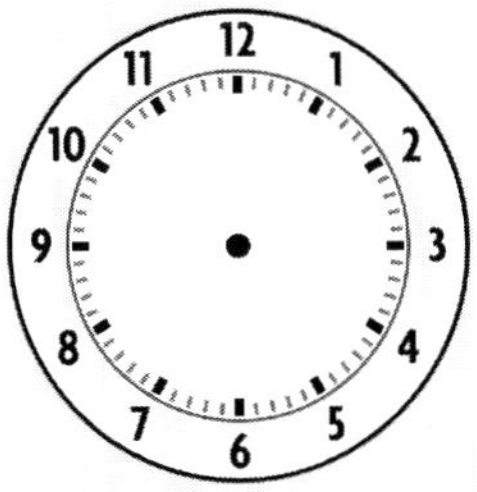
9:30

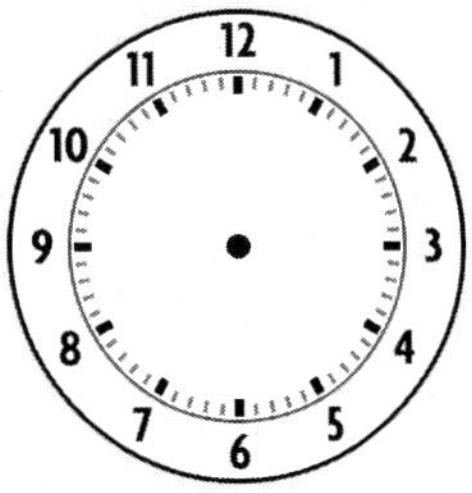
4:00

11:45

Name ______________________ Date ______________

Times Ending in :00, :15, :30, :45

Write the time in the blanks below the clocks.

   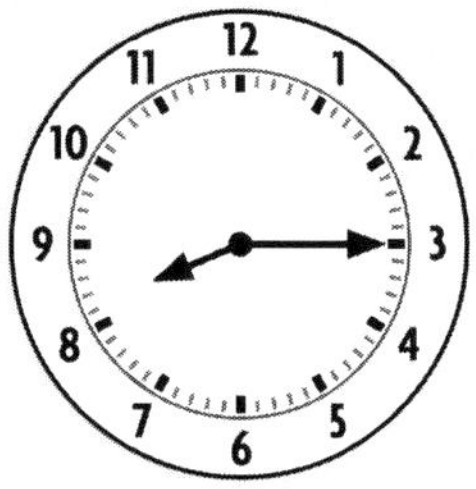

Draw the hands on the clocks to match the times.

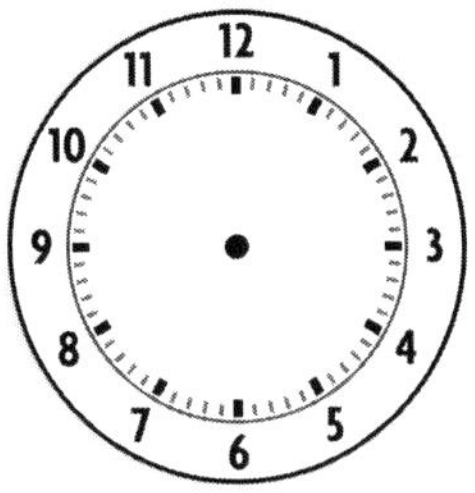
4:15

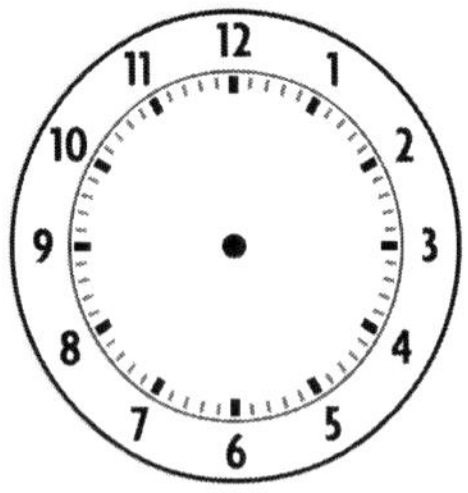
3:30

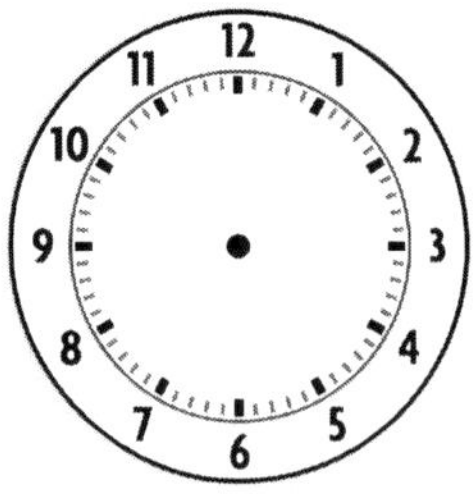
11:00

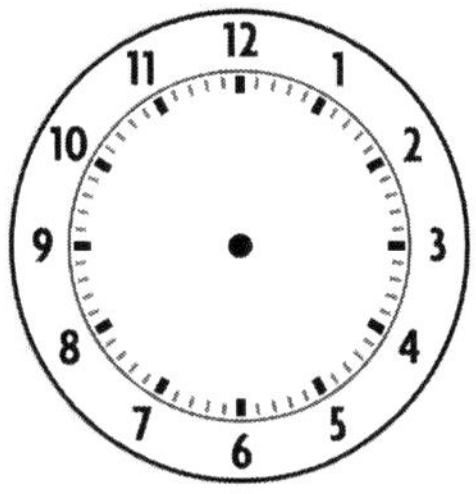
6:45

Name ______________________ Date ______________________

Times Ending in :00, :15, :30, :45

Write the time in the blanks below the clocks.

Draw the hands on the clocks to match the times.

8:30 4:30 1:45 12:15

Name ______________________ Date ______________

Times Ending in :00, :15, :30, :45

Write the time in the blanks below the clocks.

Draw the hands on the clocks to match the times.

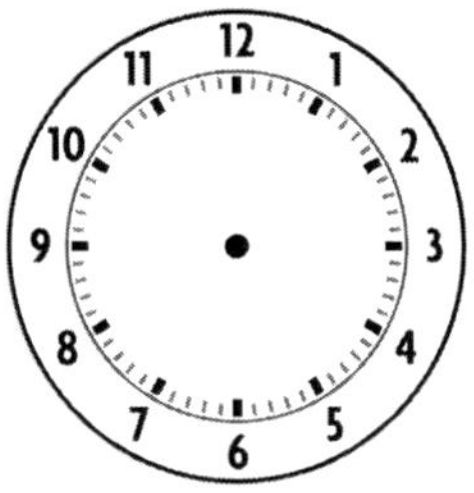
2:00

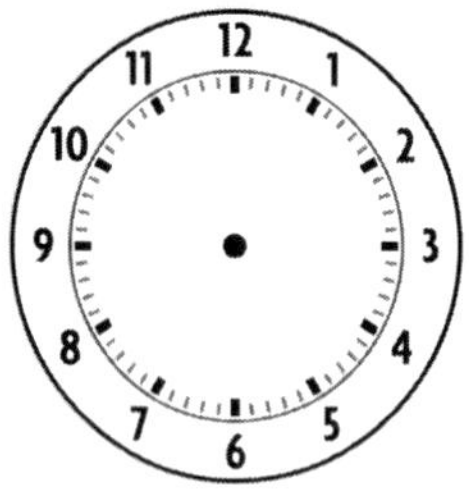
1:15

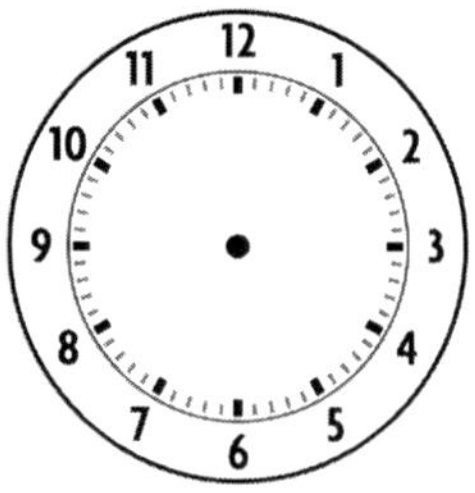
9:30

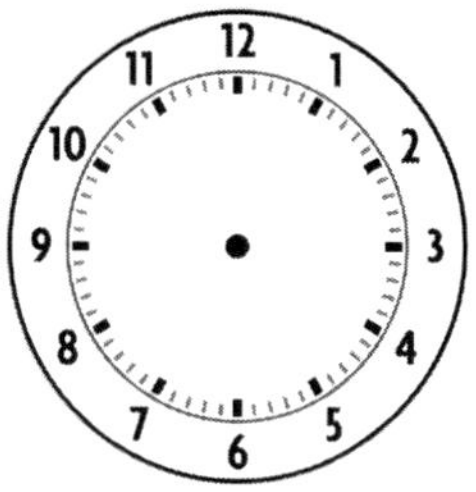
6:15

Name ______________________ Date ______________

Times Ending in :00, :15, :30, :45

Write the time below the clock showing the hands, then write the correct future time below the empty clock and draw the hands.

45 minutes later

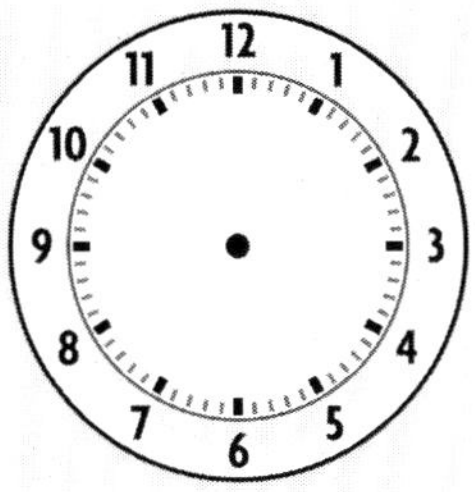

15 minutes later

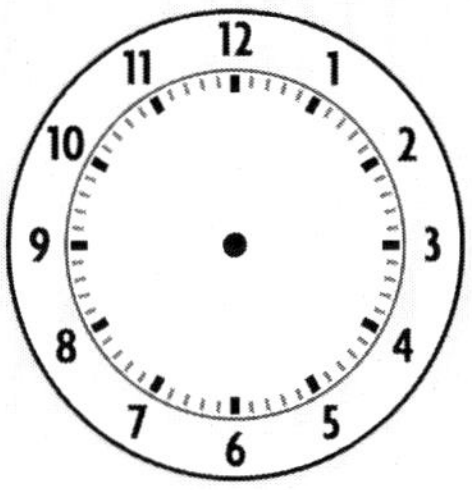

Write the time below the clock showing the hands, then write the correct past time below the empty clock and draw the hands.

45 minutes earlier

Name ______________________ Date ______________

Times Ending in :00, :15, :30, :45

Write the time below the clock showing the hands, then write the correct future time below the empty clock and draw the hands.

30 minutes later

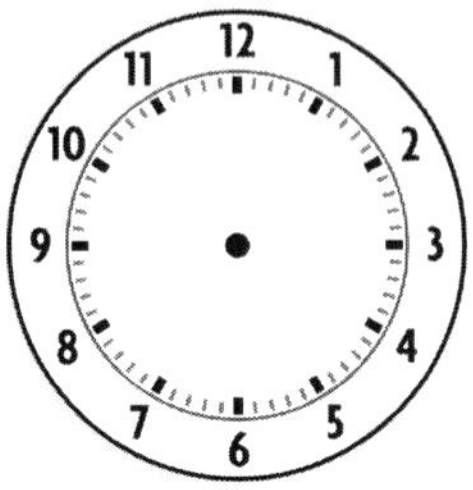

4 hours later

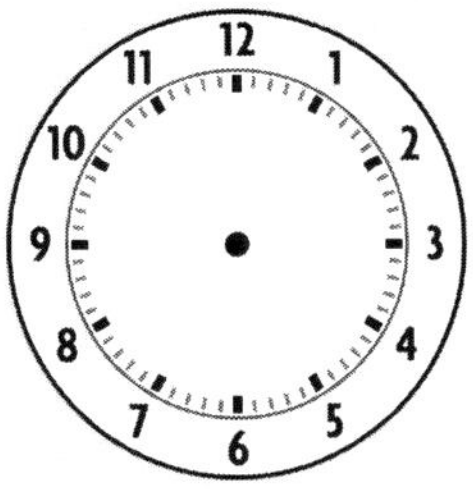

Write the time below the clock showing the hands, then write the correct past time below the empty clock and draw the hands.

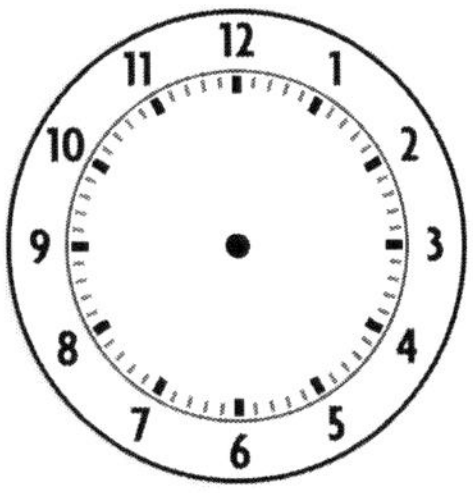

15 minutes earlier

Name ______________________ Date ______________

Times Ending in :00, :15, :30, :45

Write the time below the clock showing the hands, then write the correct future time below the empty clock and draw the hands.

15 minutes later

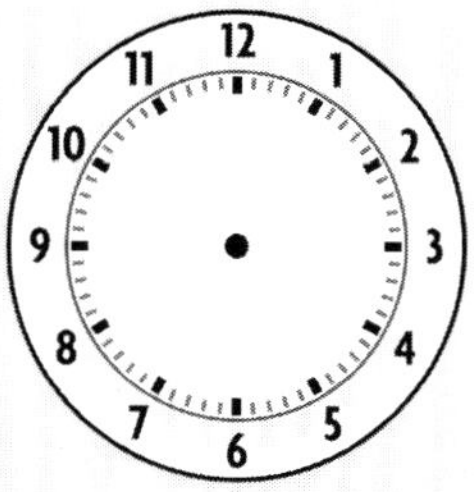

45 minutes later

Write the time below the clock showing the hands, then write the correct past time below the empty clock and draw the hands.

30 minutes earlier

Name ______________________ Date ______________

Times Ending in :00, :15, :30, :45

For each event, write the beginning and ending times under the clocks, then draw the hands on the clocks to match the correct times.

1. The bus leaves at 6:00 and arrives at school 2 hours later.

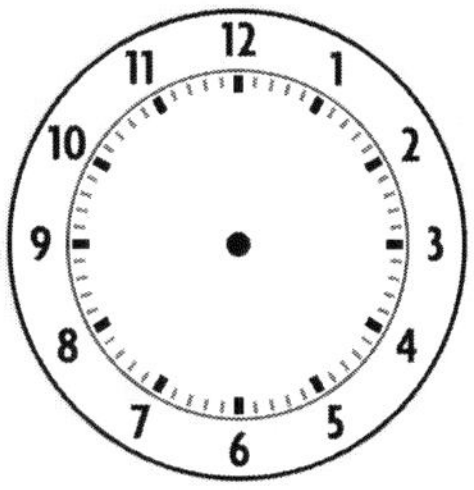

2. Dinner starts at 5:00 and lasts 45 minutes.

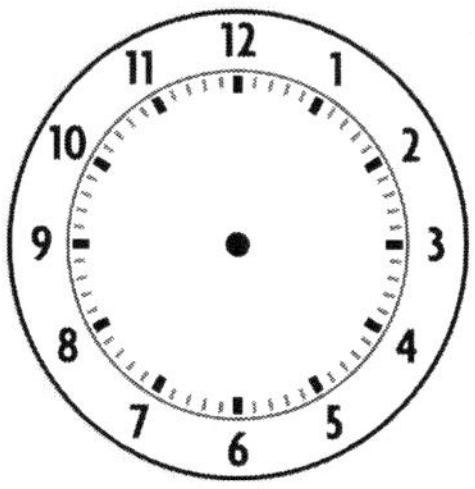
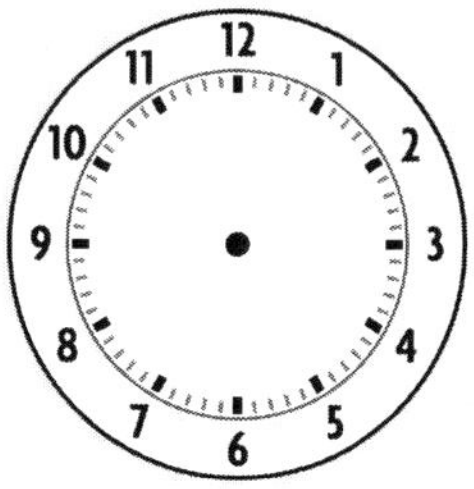

3. Hannah put a cake in the oven at 4:00 and baked it for 30 minutes.

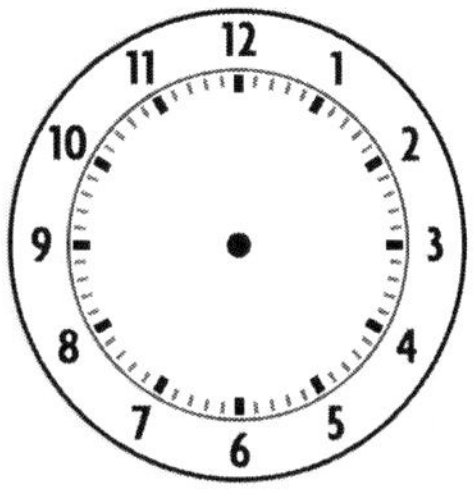

4. Roger walked his dog for 45 minutes and returned home at 3:45.

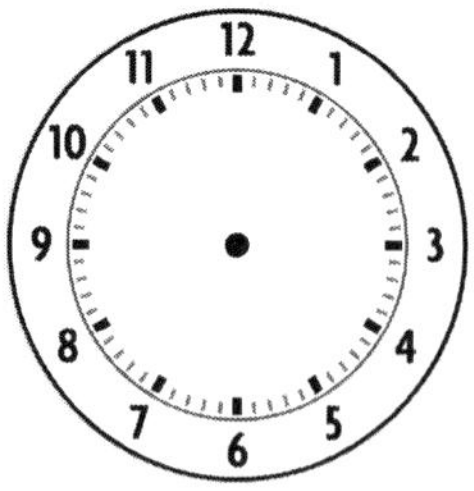
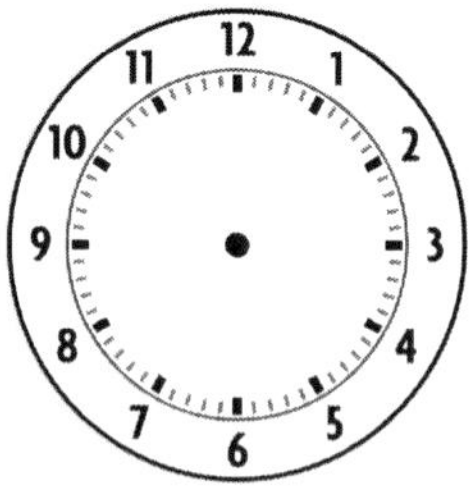

Name ______________________ Date ______________

Times Ending in :00, :15, :30, :45

For each event, write the beginning and ending times under the clocks, then draw the hands on the clocks to match the correct times.

1. Robert started reading at 11:00 and read for 15 minutes.

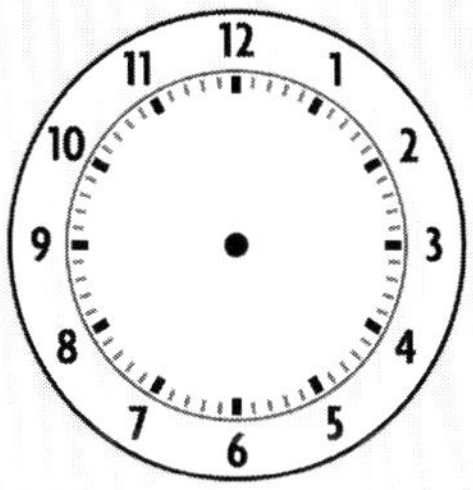

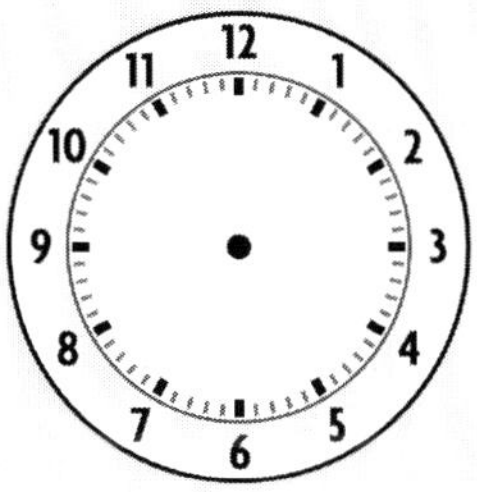

2. Jasmine started riding her bike at 3:30 and finished 1 hour later.

3. Percy practiced gymnastics for 1 hour. He started at 2:45.

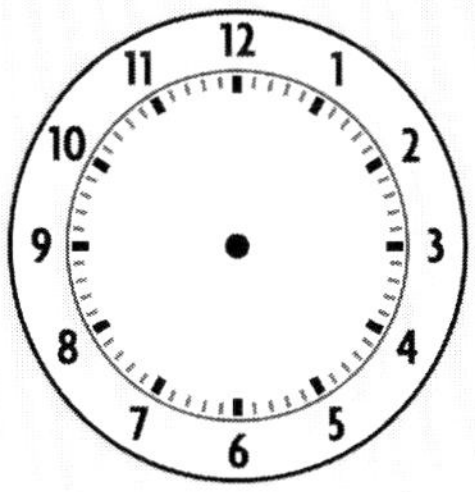

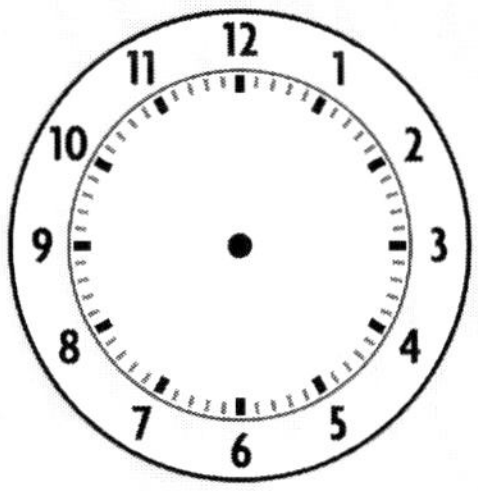

4. Shawna painted for 45 minutes and finished at 7:45.

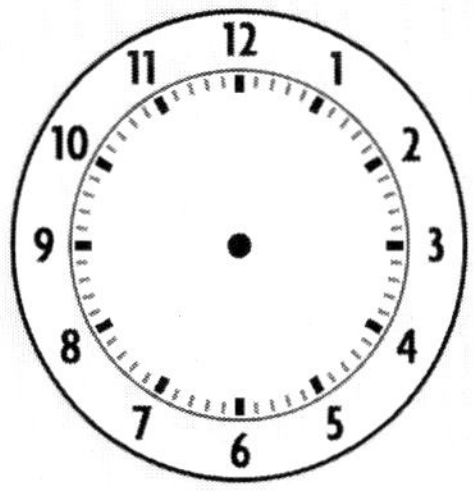

Name ______________________________ Date ____________________

Times Ending in :00, :15, :30, :45

For each event, write the beginning and ending times under the clocks, then draw the hands on the clocks to match the correct times.

1. The movie started at 8:00 and ended 2 hours later.

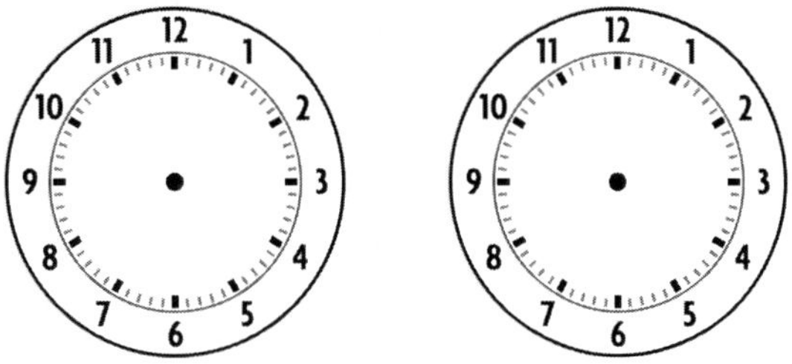

2. Jeremy began eating breakfast at 6:00 and finished 15 minutes later.

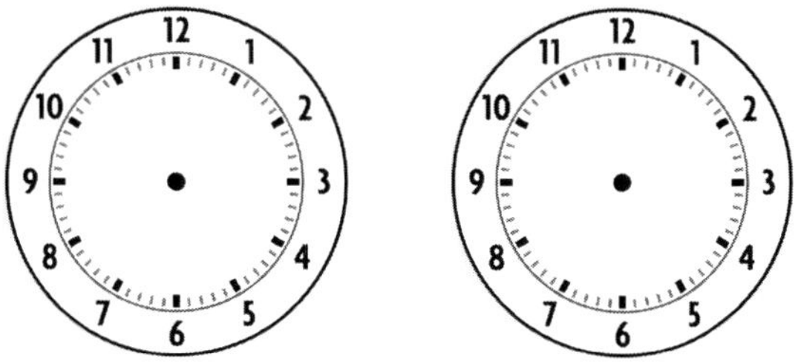

3. The parade started at 8:30 and stopped 2 hours later.

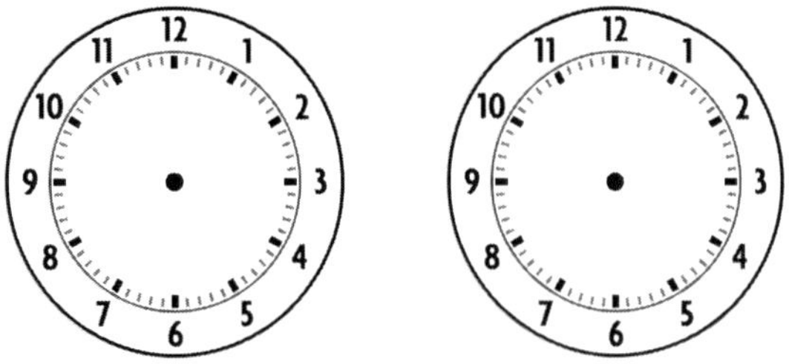

4. The children swam for 45 minutes. They stopped at 1:45.

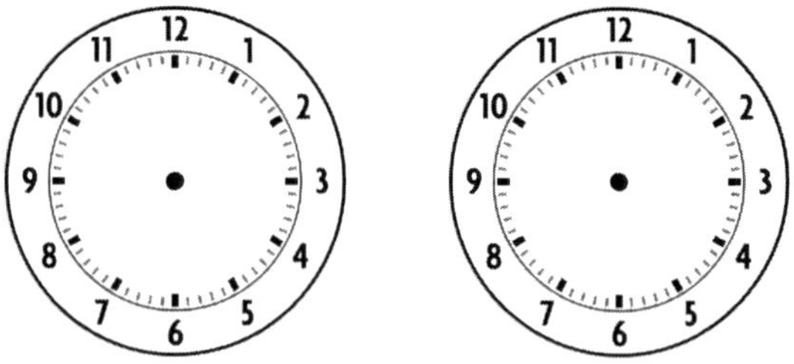

SECTION 6

Times Ending in 5-Minute Marks **:00, :05, :10, :15, :20, ...**

Worksheet Types

Match Clocks to Times
Read the Time, Write the Time, Draw the Hands
Future Time, Past Time
Word Problems

Name ______________________________ Date ______________________

Times Ending in 5-Minute Marks (:00, :05, :10, :15, ...)

Draw lines to match the clocks to the correct times.

4:50

1:25

7:05

12:40

4:35

Name ______________________ Date ______________

Times Ending in 5-Minute Marks (:00, :05, :10, :15, ...)

Draw lines to match the clocks to the correct times.

10:20

11:35

7:25

6:40

5:10

Name ______________________________ Date ____________________

Times Ending in 5-Minute Marks (:00, :05, :10, :15, ...)

Write the time in the blanks below the clocks.

Draw the hands on the clocks to match the times.

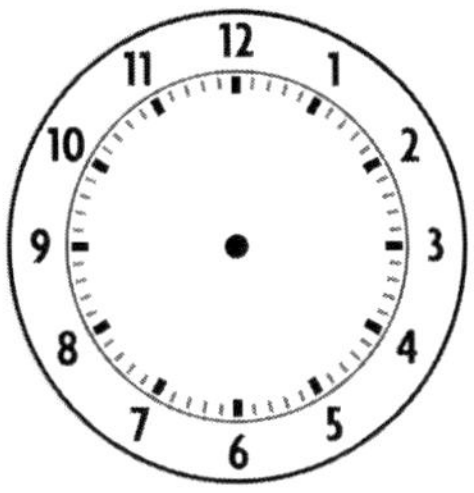
1:10

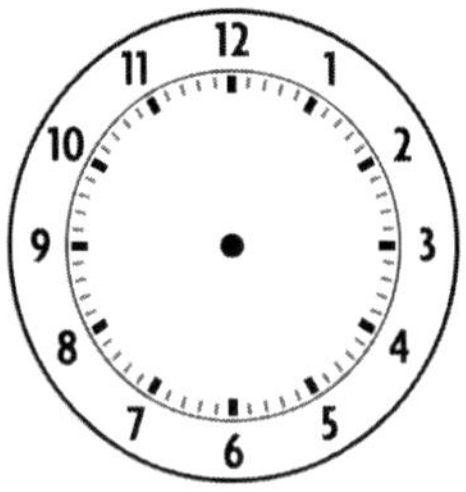
5:40

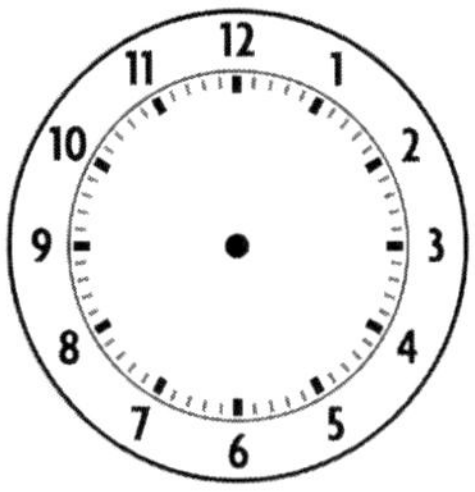
6:05

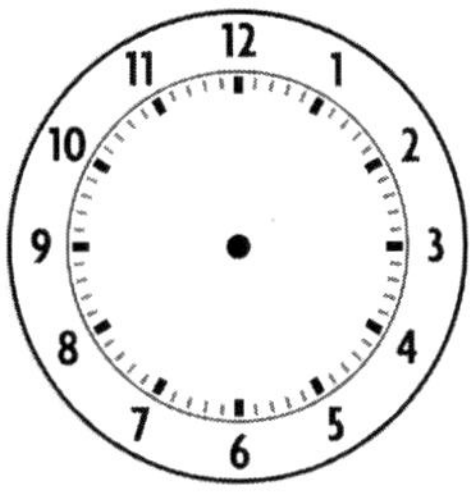
2:35

Name ______________________________ Date ________________

Times Ending in 5-Minute Marks (:00, :05, :10, :15, ...)

Write the time in the blanks below the clocks.

Draw the hands on the clocks to match the times.

2:15	10:35	6:20	8:25

Name ______________________ Date ______________

Times Ending in 5-Minute Marks (:00, :05, :10, :15, ...)

Write the time in the blanks below the clocks.

Draw the hands on the clocks to match the times.

8:35	9:40	6:50	3:20

Name ______________________ Date ______________

Times Ending in 5-Minute Marks (:00, :05, :10, :15, ...)

Write the time in the blanks below the clocks.

Draw the hands on the clocks to match the times.

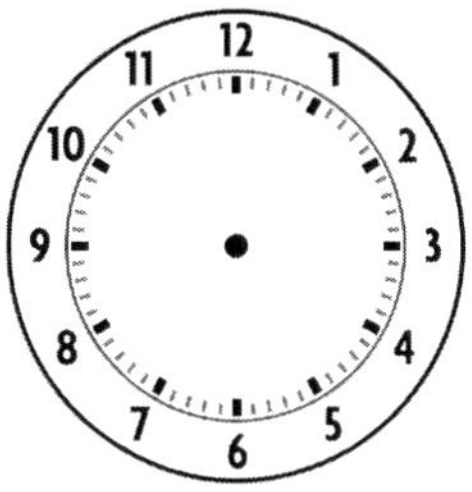
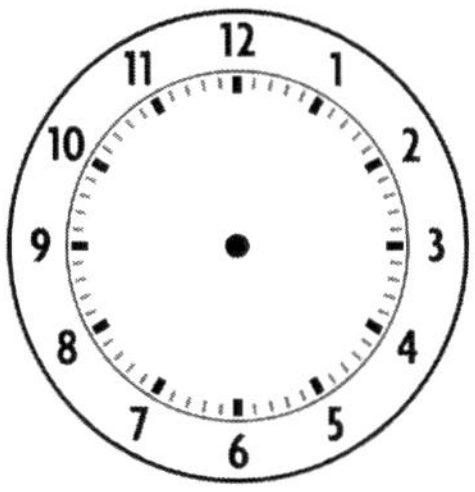
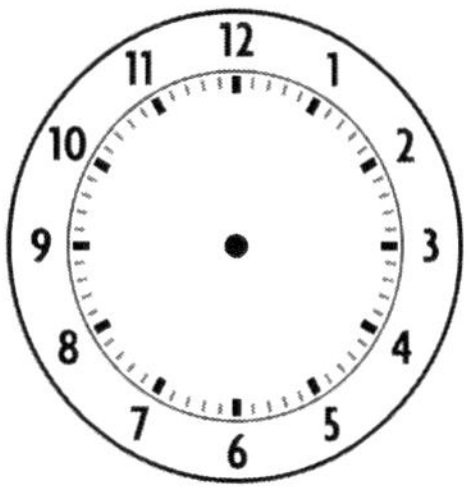
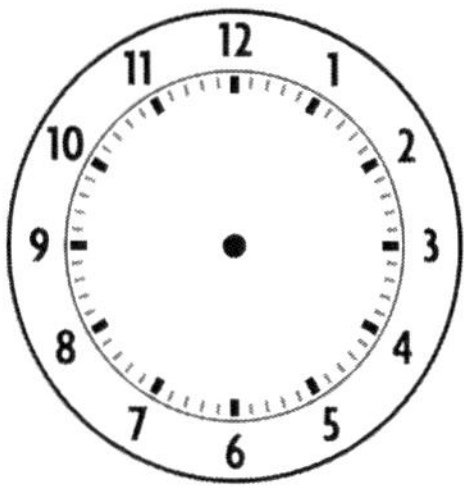

11:50 | 3:35 | 1:05 | 10:40

Name ______________________ Date ______________

Times Ending in 5-Minute Marks (:00, :05, :10, :15, ...)

Write the time in the blanks below the clocks.

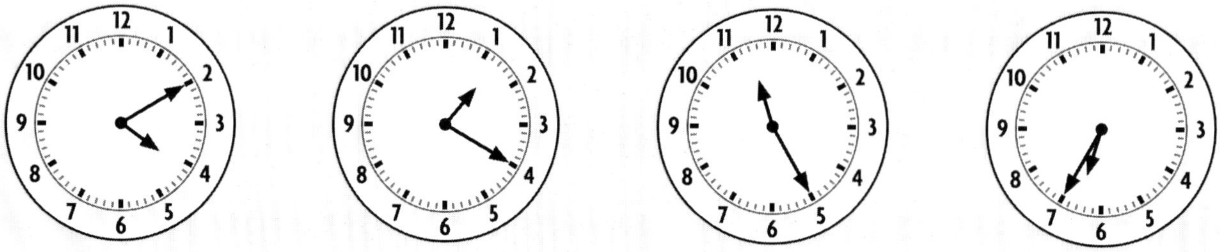

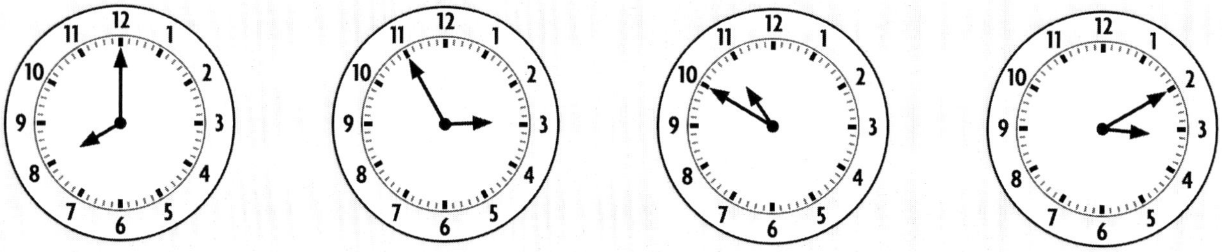

Draw the hands on the clocks to match the times.

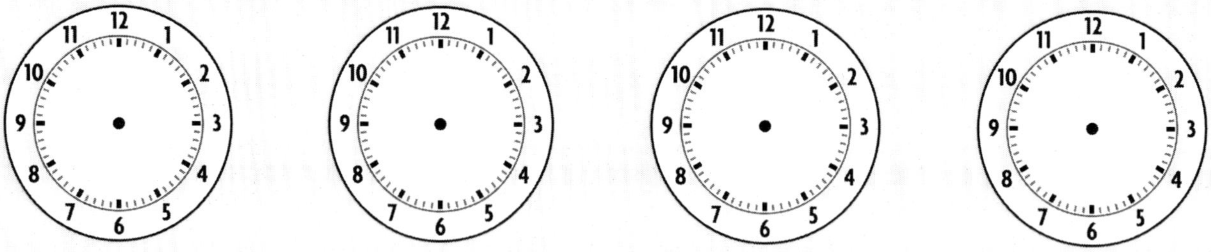

10:25 11:40 9:55 4:25

Name ______________________ Date ______________

Times Ending in 5-Minute Marks (:00, :05, :10, :15, ...)

Write the time in the blanks below the clocks.

Draw the hands on the clocks to match the times.

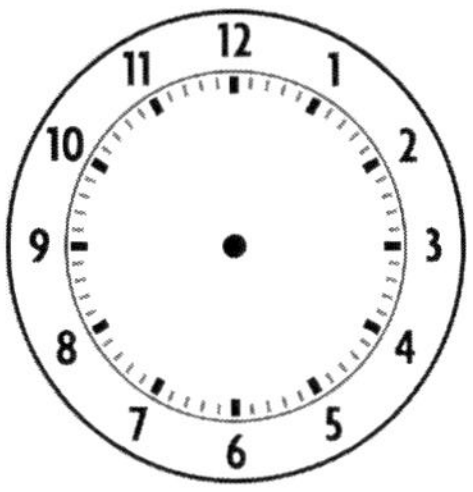
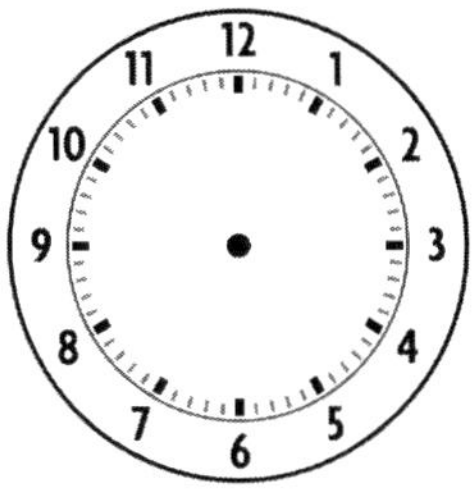
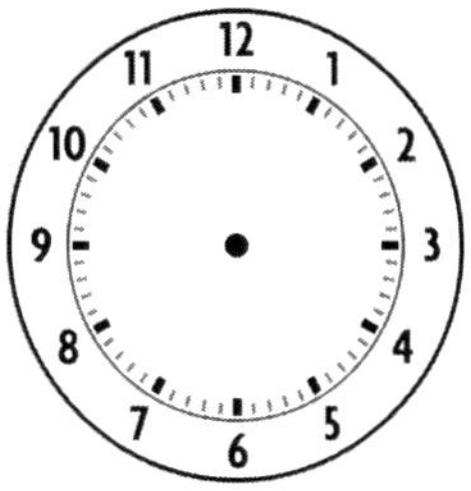
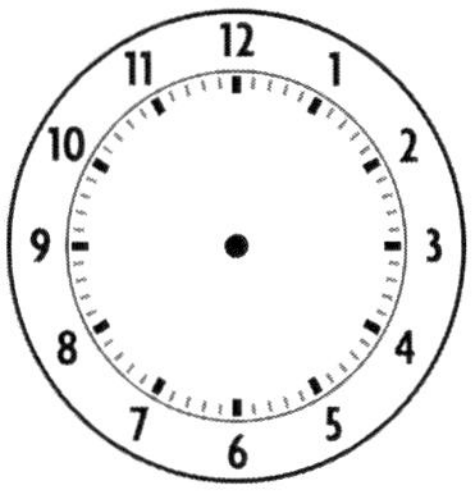

11:05	3:50	9:20	1:35

Name ______________________________ Date ____________________

Times Ending in 5-Minute Marks (:00, :05, :10, :15, ...)

Write the time in the blanks below the clocks.

Draw the hands on the clocks to match the times.

 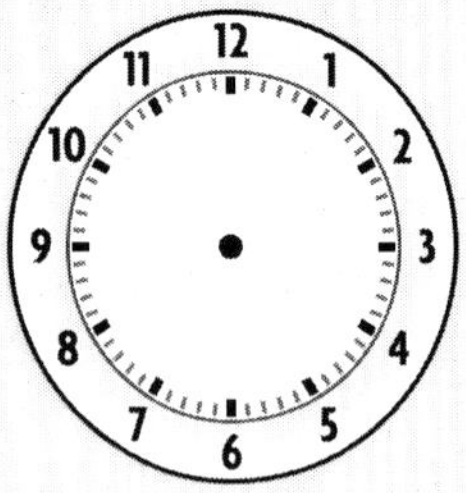  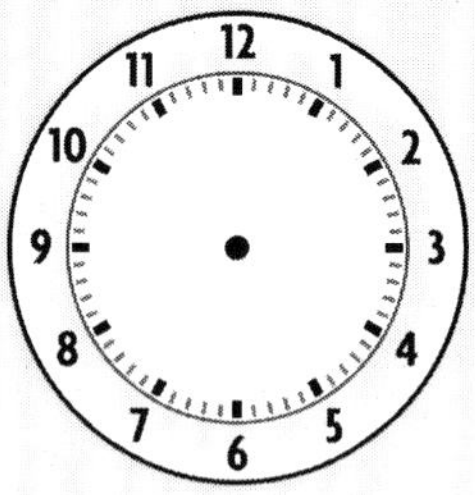

8:10 4:20 7:10 12:25

Name ________________________________ Date ____________________

Times Ending in 5-Minute Marks (:00, :05, :10, :15, ...)

Write the time in the blanks below the clocks.

Draw the hands on the clocks to match the times.

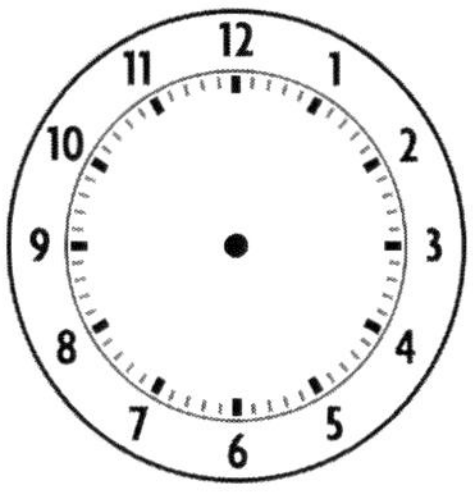
3:55

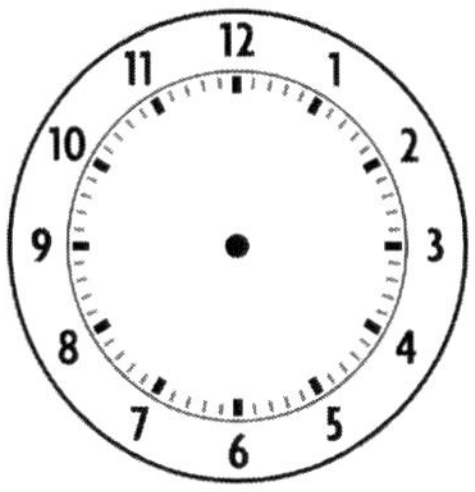
2:25

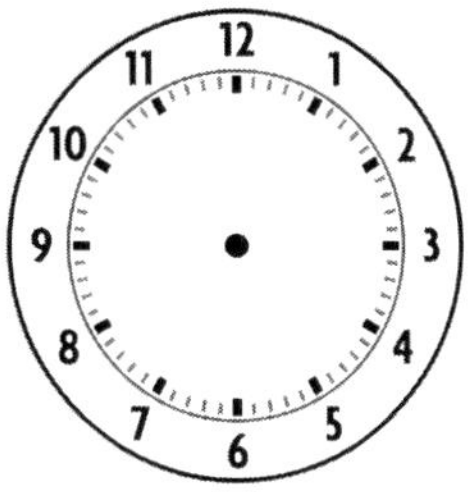
4:05

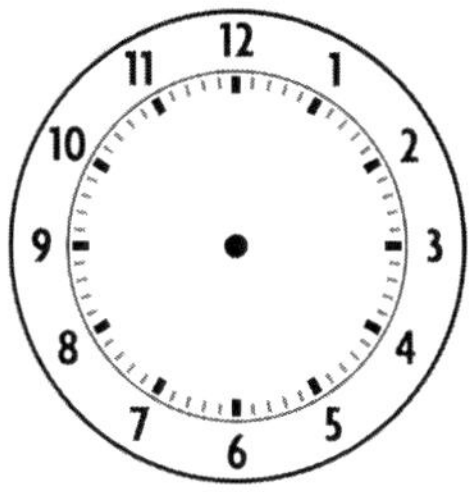
10:10

Name ______________________ Date ______________

Times Ending in 5-Minute Marks (:00, :05, :10, :15, ...)

Write the time below the clock showing the hands, then write the correct future time below the empty clock and draw the hands.

5 minutes later

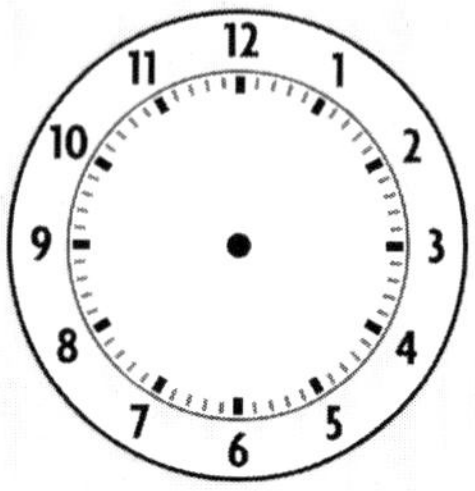

10 minutes later

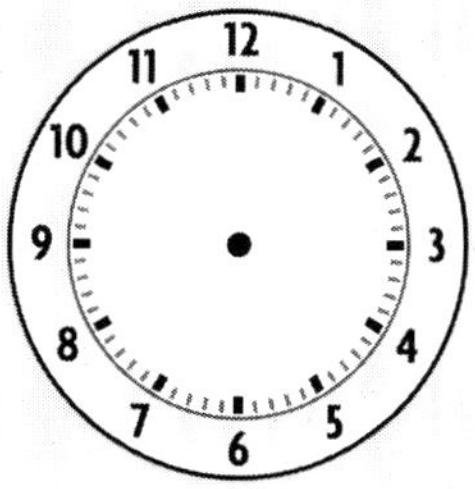

Write the time below the clock showing the hands, then write the correct past time below the empty clock and draw the hands.

5 minutes earlier

Name ____________________ Date ____________________

Times Ending in 5-Minute Marks (:00, :05, :10, :15, ...)

Write the time below the clock showing the hands, then write the correct future time below the empty clock and draw the hands.

35 minutes later

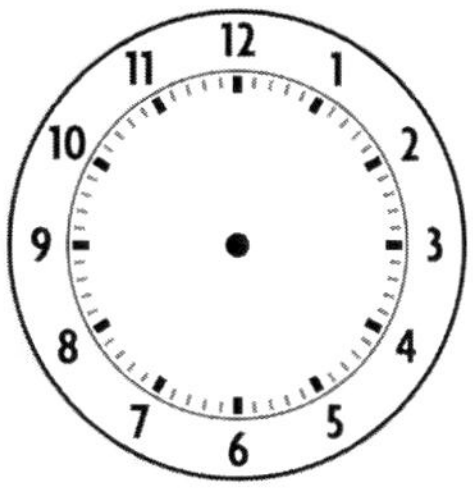

5 minutes later

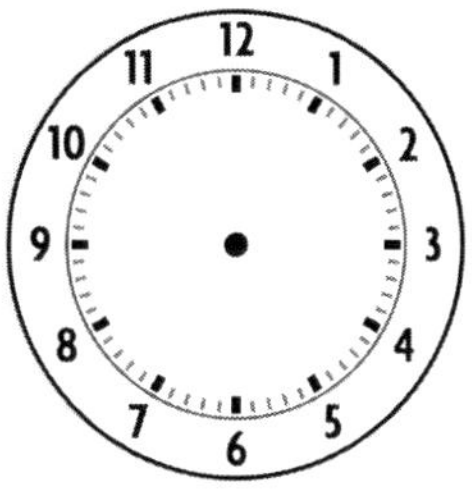

Write the time below the clock showing the hands, then write the correct past time below the empty clock and draw the hands.

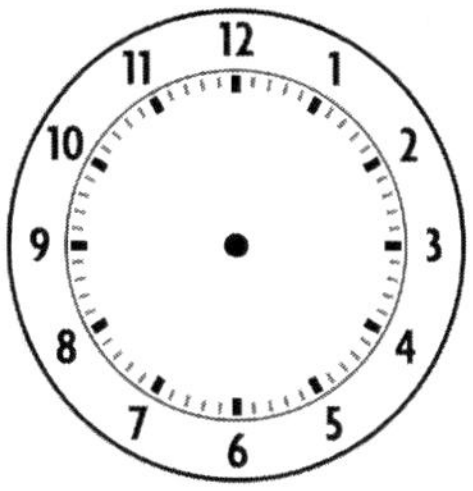

10 minutes earlier

Name ______________________________ Date ____________________

Times Ending in 5-Minute Marks (:00, :05, :10, :15, ...)

Write the time below the clock showing the hands, then write the correct future time below the empty clock and draw the hands.

5 minutes later

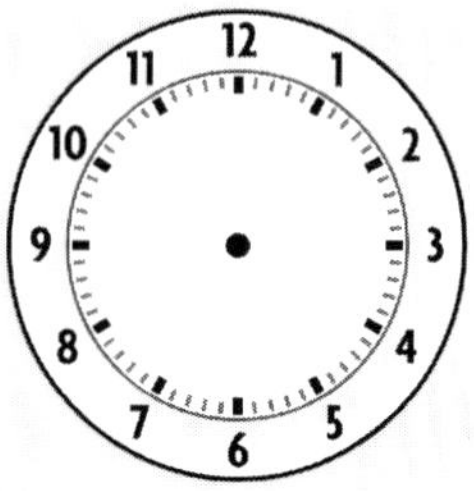

20 minutes later

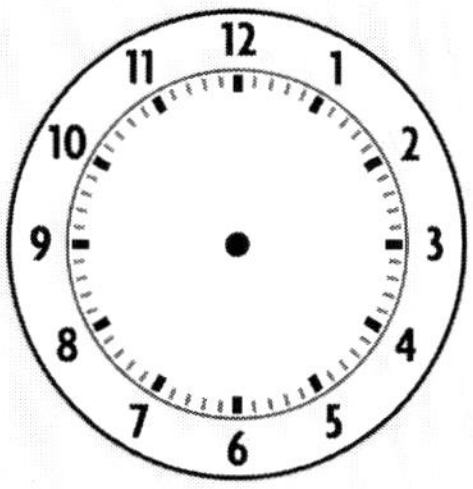

Write the time below the clock showing the hands, then write the correct past time below the empty clock and draw the hands.

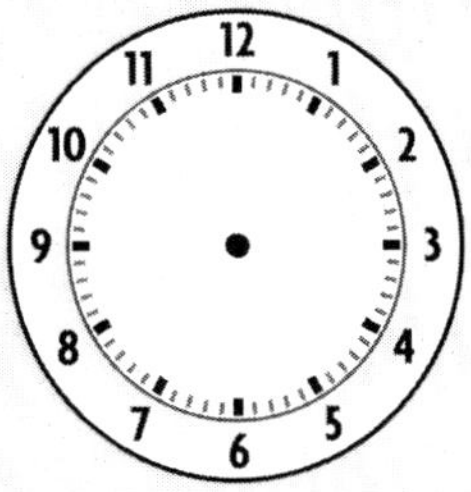

10 minutes earlier

Name ______________________________ Date ____________________

Times Ending in 5-Minute Marks (:00, :05, :10, :15, ...)

For each event, write the beginning and ending times under the clocks, then draw the hands on the clocks to match the correct times.

1. Misha began eating a sandwich at 10:20 and finished eating in 5 minutes.

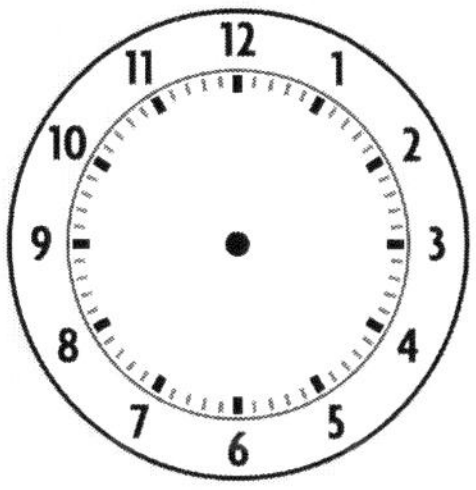

2. Jeremy practiced batting at 2:30 and stopped 10 minutes later.

3. Dell spent 1 hour building a LEGO fort. He started at 4:35.

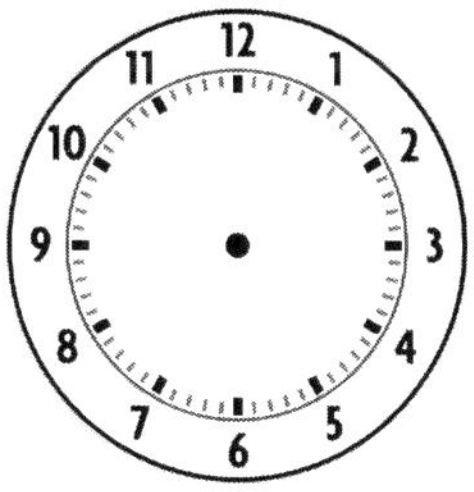
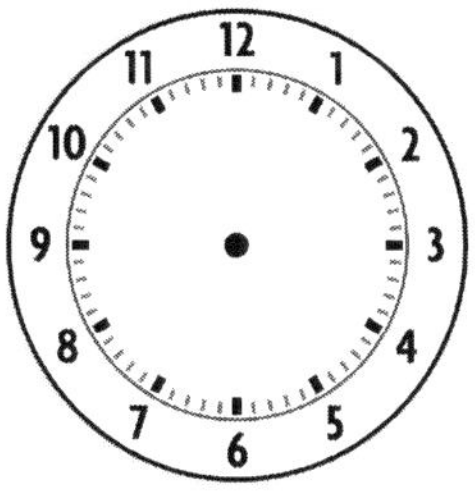

4. Jane practiced jumping for 5 minutes and stopped at 12:25.

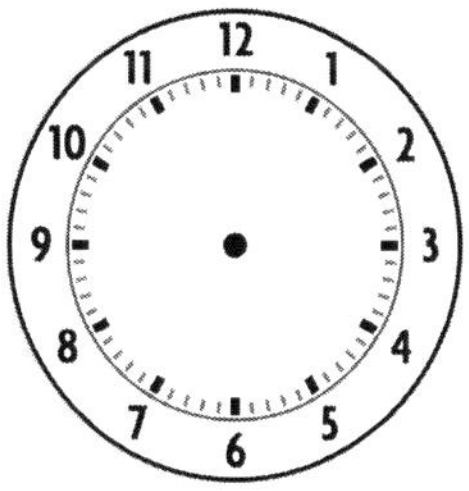
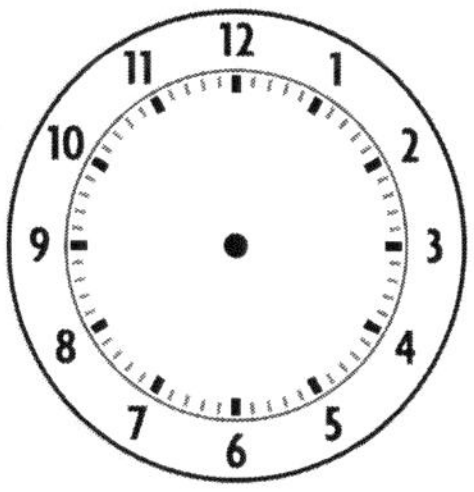

Name ______________________ Date ______________

Times Ending in 5-Minute Marks (:00, :05, :10, :15, ...)

For each event, write the beginning and ending times under the clocks, then draw the hands on the clocks to match the correct times.

1. The concert began at 6:40 and was over in 2 hours.

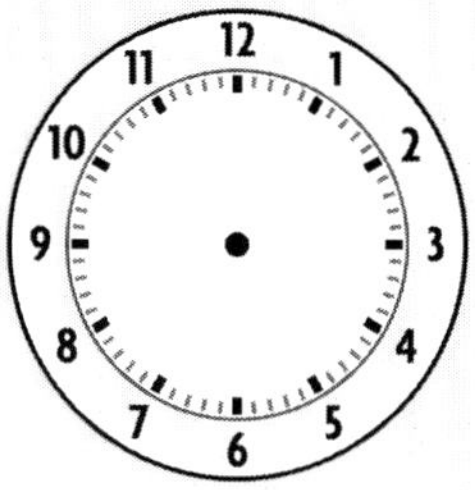

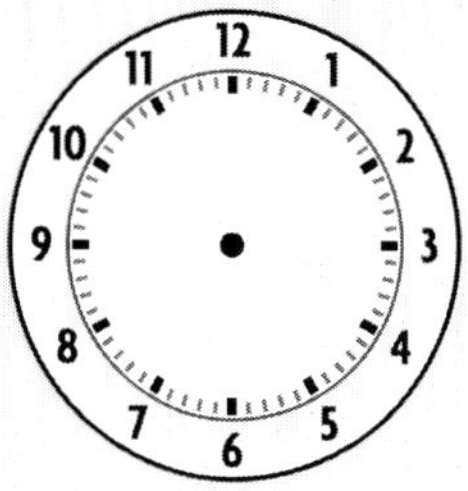

2. Grista started her worksheet at 5:45 and finished in 5 minutes.

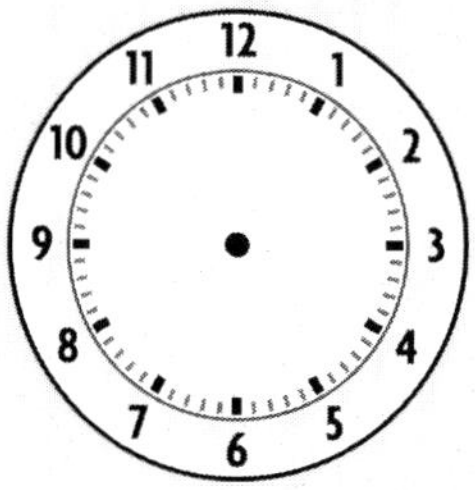

3. Abby ran for 10 minutes. She started at 9:20.

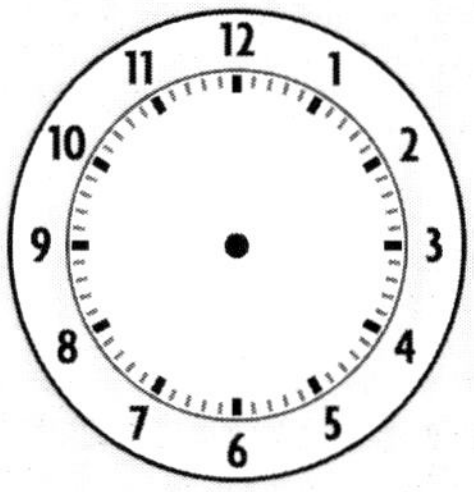

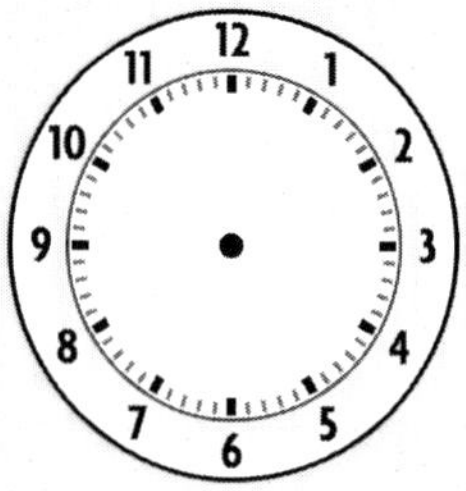

4. The fan stopped spinning at 7:55. It was on for 5 minutes.

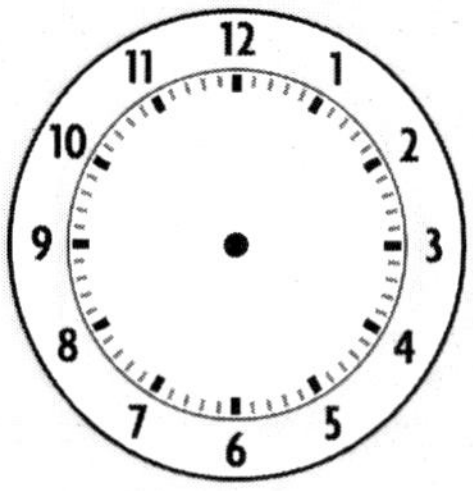

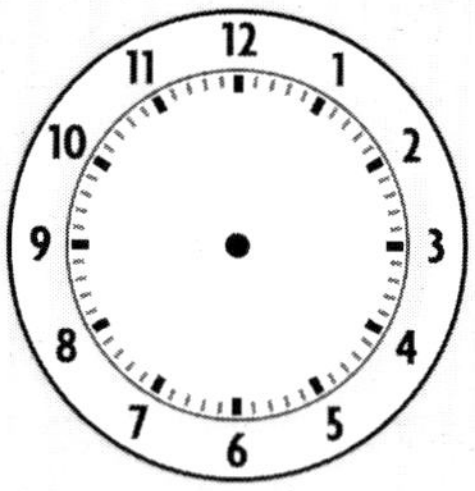

Name ______________________ Date ______________

Times Ending in 5-Minute Marks (:00, :05, :10, :15, ...)

For each event, write the beginning and ending times under the clocks, then draw the hands on the clocks to match the correct times.

1. Hugo left his house 1:20 and walked for 5 minutes to his friend's house.

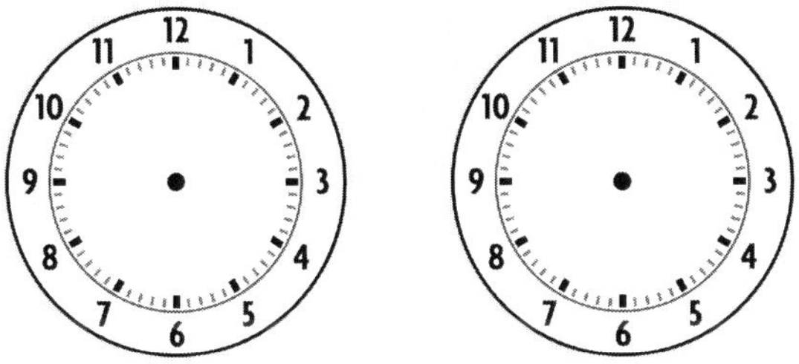

2. Shannon started cleaning her room at 4:00 and finished 35 minutes later.

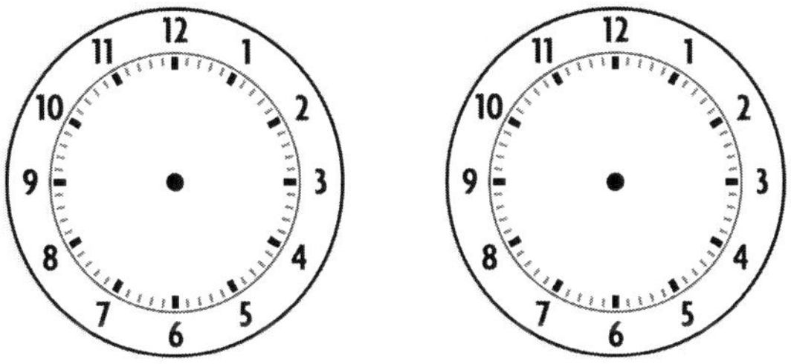

3. Rita played video games with her little brother at 11:50 and they played for 1 hour.

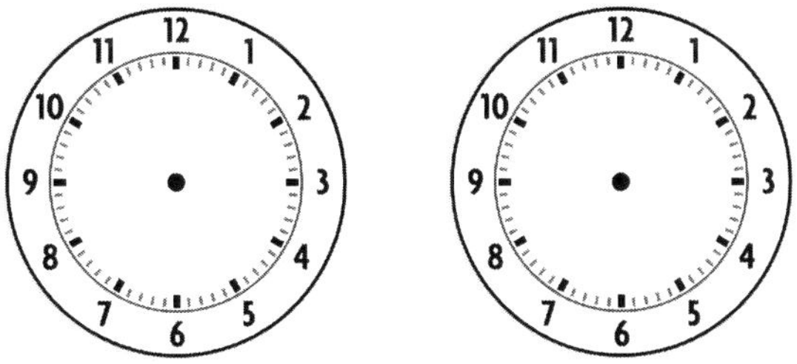

4. Jillian's cookies were finished baking at 5:30. They were in the oven for 10 minutes.

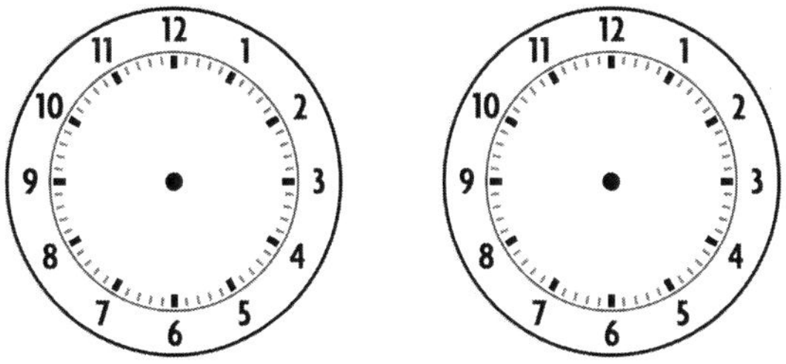

Name ______________________________ Date ____________________

Times Ending in 5-Minute Marks (:00, :05, :10, :15, ...)

For each event, write the beginning and ending times under the clocks, then draw the hands on the clocks to match the correct times.

1. Isabelle rode her horse for 1 hour. She started at 4:40.

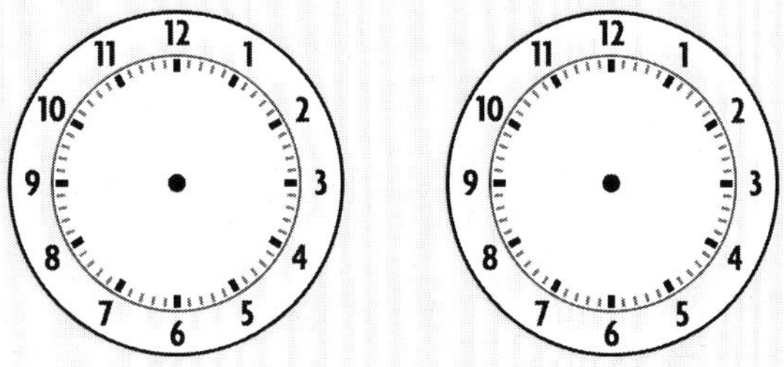

2. Mark started eating an apple at 11:20 and finished in 5 minutes.

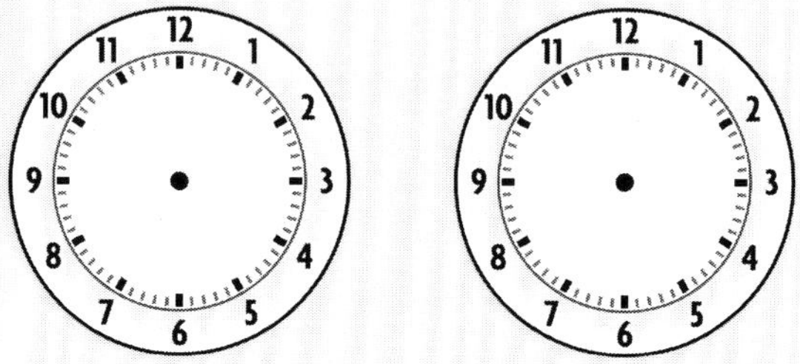

3. Erin brushed her hair for 10 minutes. She started at 8:10.

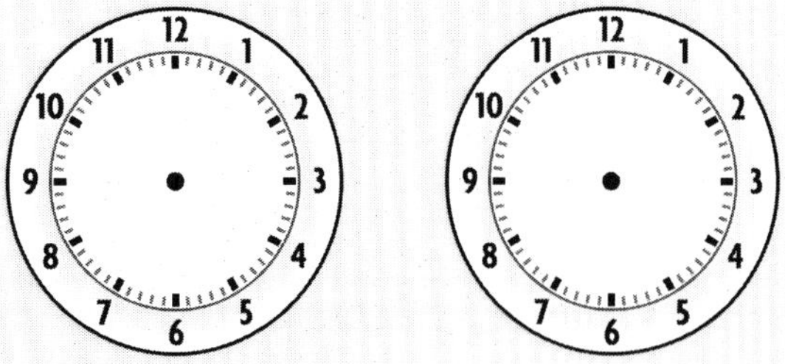

4. The popcorn finished popping at 6:15. It popped for 5 minutes.

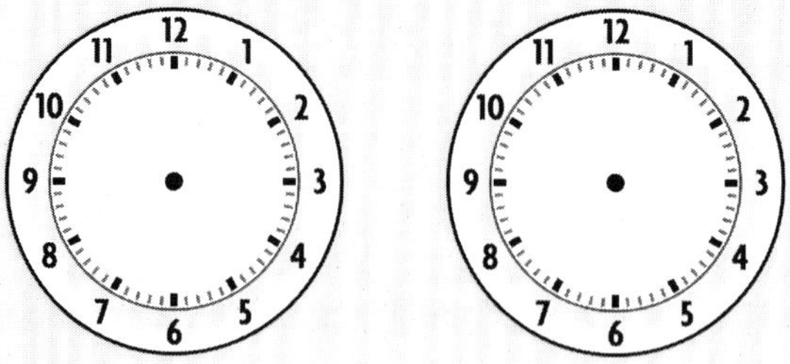

SECTION

7

Times Ending in 1-Minute Marks **:00, :01, :02, :03, :04, ...**

Worksheet Types

Match Clocks to Times
Read the Time, Write the Time, Draw the Hands
Future Time, Past Time
Word Problems

Name ______________________________ Date ____________________

Times Ending in 1-Minute Marks (:00, :01, :02, :03, ...)

Draw lines to match the clocks to the correct times.

11:49

3:03

7:41

9:52

7:08

Name ______________________________ Date ______________________

Times Ending in 1-Minute Marks (:00, :01, :02, :03, ...)

Draw lines to match the clocks to the correct times.

5:21

12:19

6:52

8:41

2:34

Name ________________________ Date ______________

Times Ending in 1-Minute Marks (:00, :01, :02, :03, ...)

Write the time in the blanks below the clocks.

Draw the hands on the clocks to match the times.

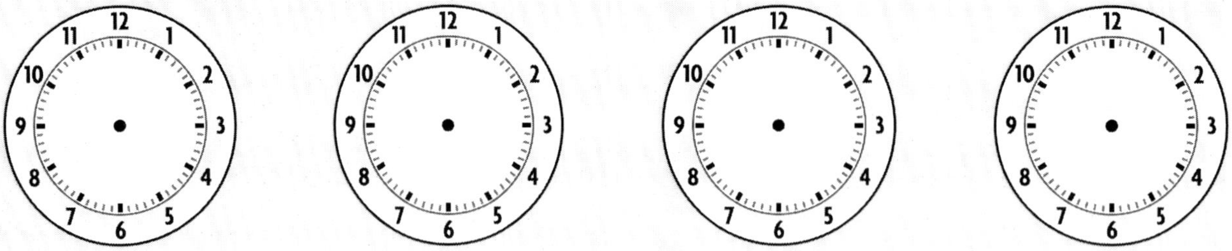

6:29	1:33	11:07	1:11

Name ______________________________ Date ____________________

Times Ending in 1-Minute Marks (:00, :01, :02, :03, ...)

Write the time in the blanks below the clocks.

Draw the hands on the clocks to match the times.

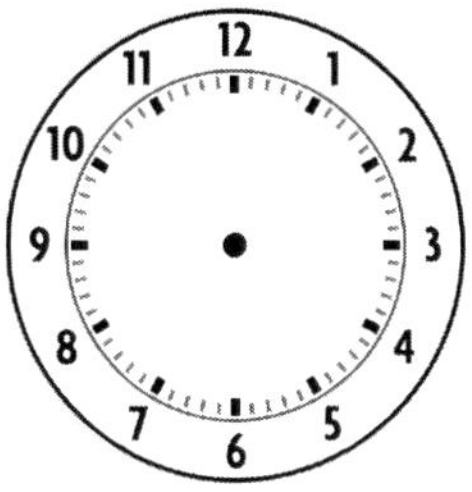
7:10

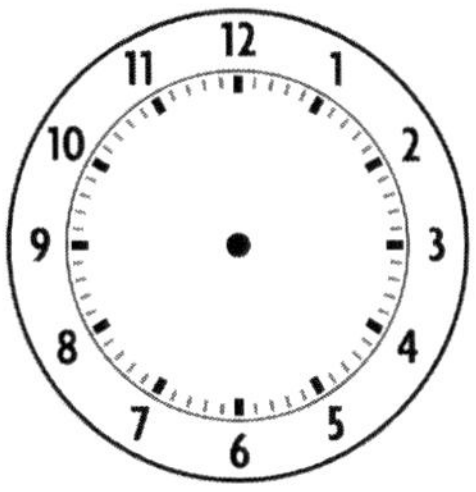
4:57

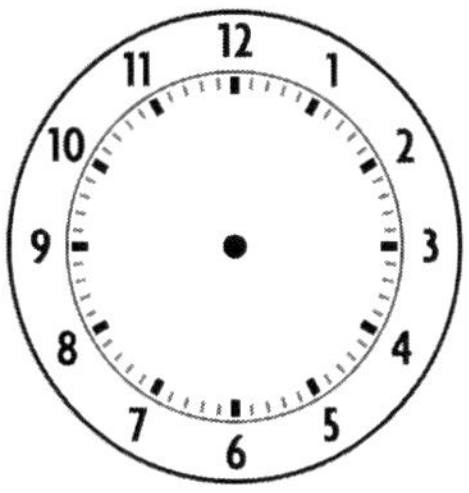
5:33

2:56

Name ______________________ Date ______________

Times Ending in 1-Minute Marks (:00, :01, :02, :03, ...)

Write the time in the blanks below the clocks.

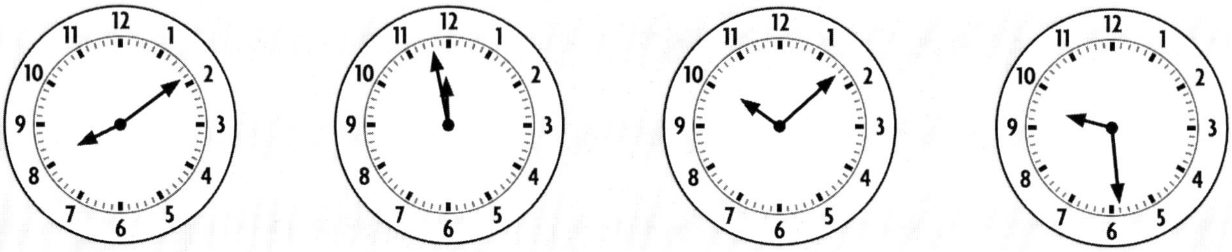

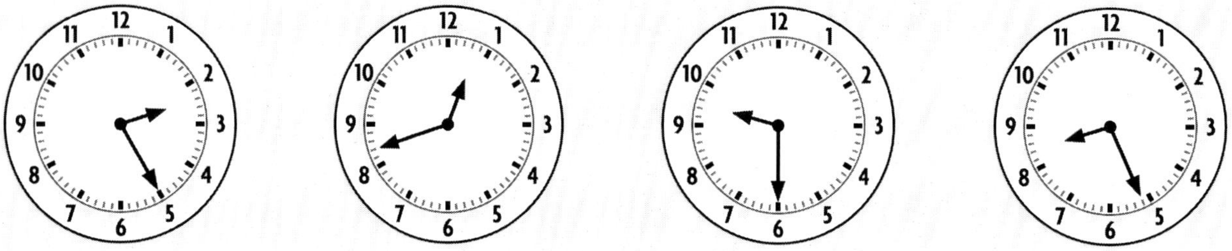

Draw the hands on the clocks to match the times.

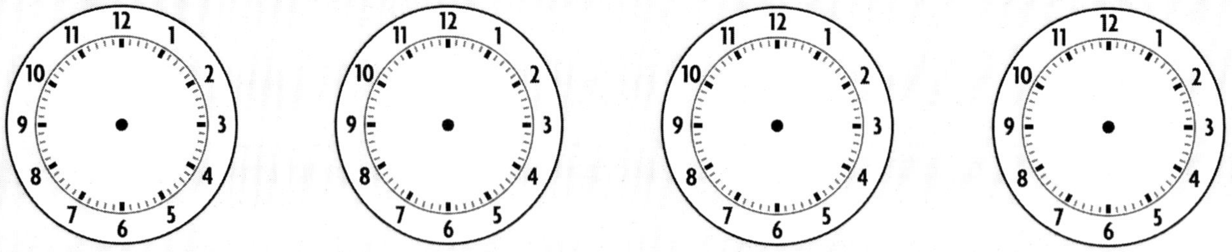

12:58 | 5:14 | 10:19 | 9:45

Name ______________________________ Date ____________________

Times Ending in 1-Minute Marks (:00, :01, :02, :03, ...)

Write the time in the blanks below the clocks.

Draw the hands on the clocks to match the times.

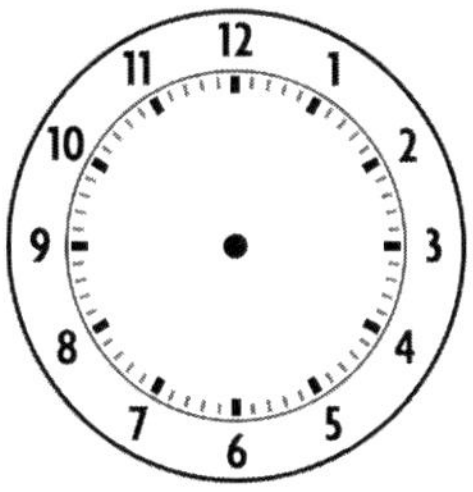

4:47

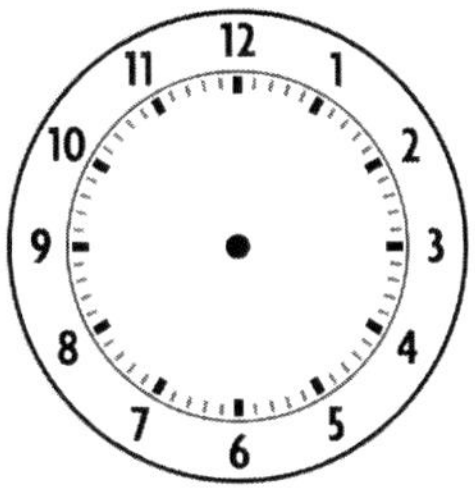

2:58

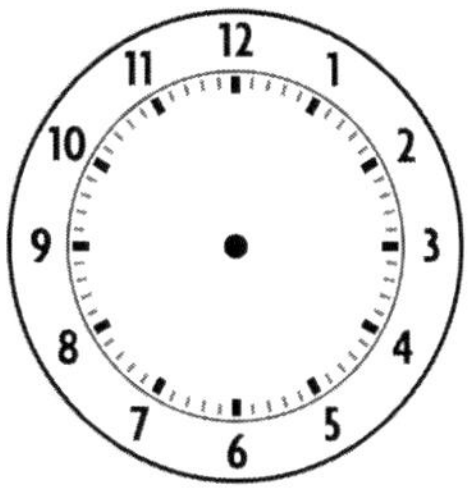

4:35

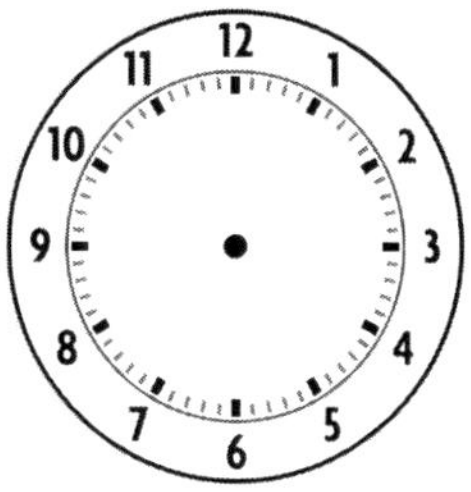

1:39

Name ______________________________ Date ____________________

Times Ending in 1-Minute Marks (:00, :01, :02, :03, ...)

Write the time in the blanks below the clocks.

Draw the hands on the clocks to match the times.

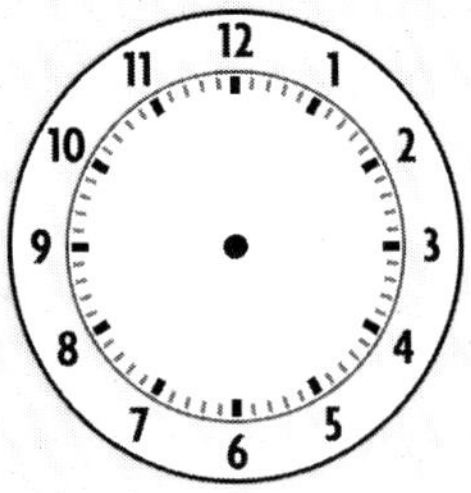
11:18

5:28

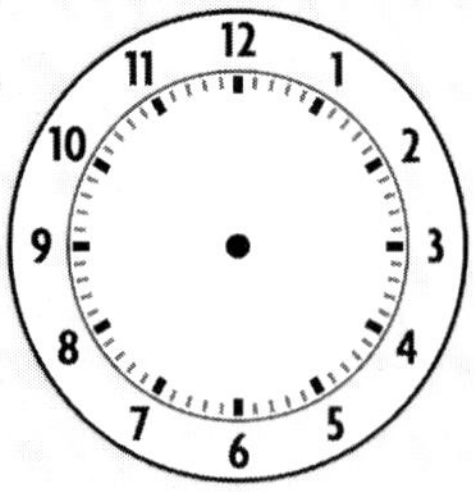
7:12

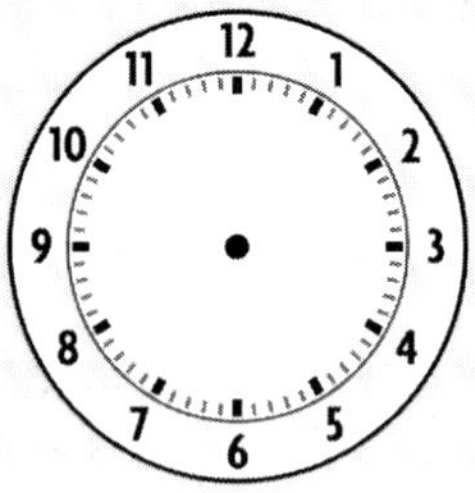
12:47

Name ______________________________ Date ____________________

Times Ending in 1-Minute Marks (:00, :01, :02, :03, ...)

Write the time in the blanks below the clocks.

Draw the hands on the clocks to match the times.

7:07

1:57

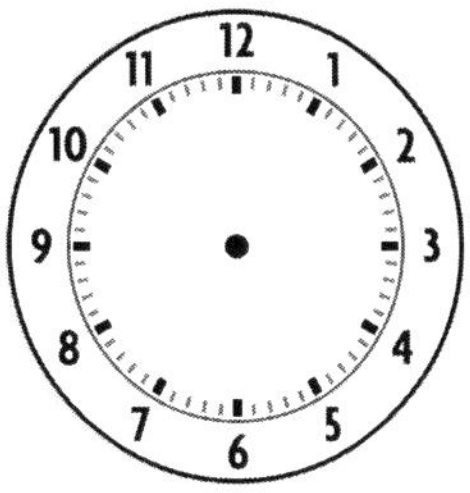

9:04

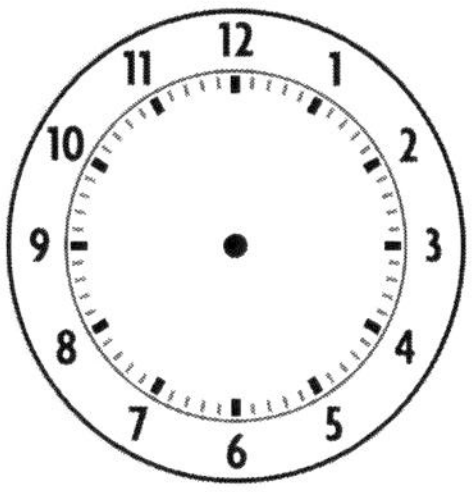

2:52

Name ______________________ Date ______________

Times Ending in 1-Minute Marks (:00, :01, :02, :03, ...)

Write the time in the blanks below the clocks.

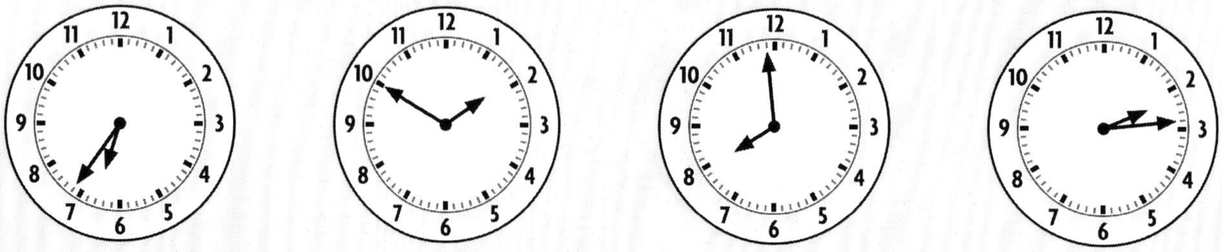

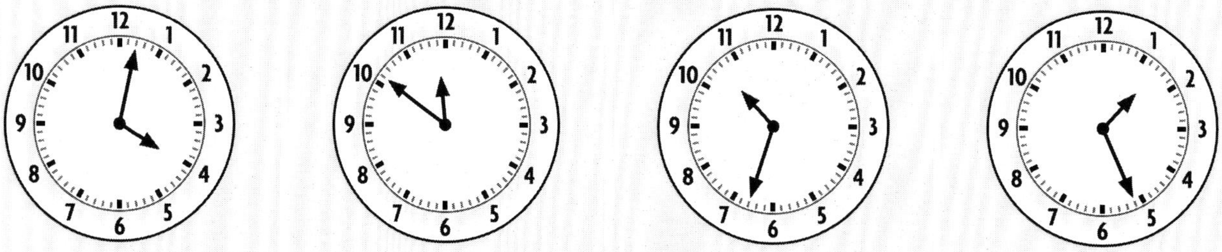

Draw the hands on the clocks to match the times.

6:49 2:23 5:18 8:22

Name ______________________________ Date ____________________

Times Ending in 1-Minute Marks (:00, :01, :02, :03, ...)

Write the time in the blanks below the clocks.

Draw the hands on the clocks to match the times.

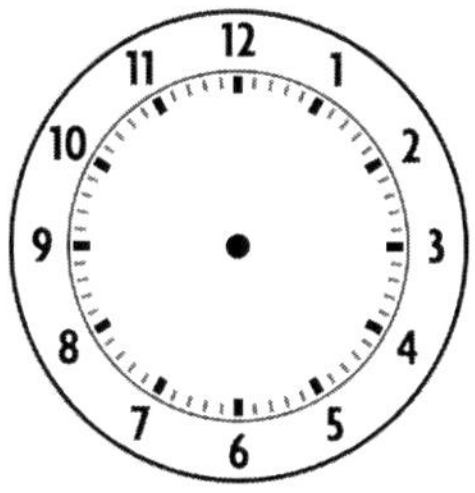
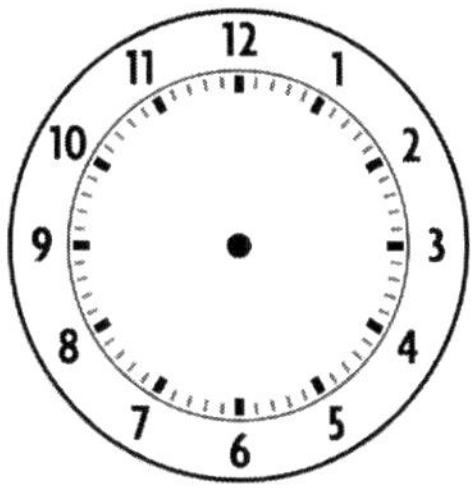
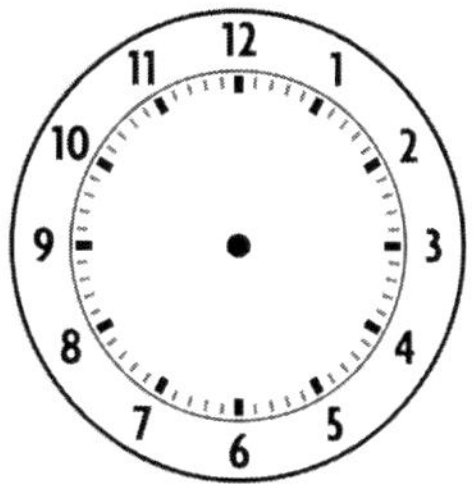
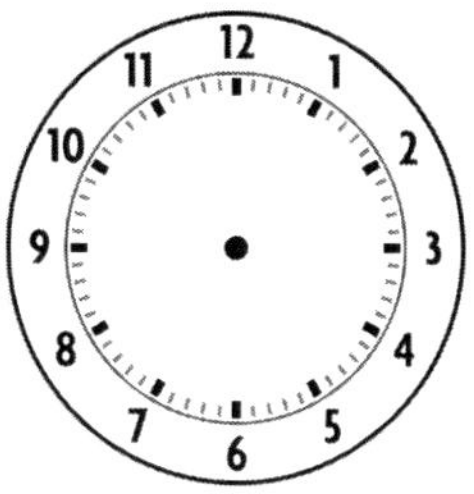

12:31 | 8:42 | 10:39 | 3:57

Name ______________________________ Date ____________________

Times Ending in 1-Minute Marks (:00, :01, :02, :03, ...)

Write the time below the clock showing the hands, then write the correct future time below the empty clock and draw the hands.

1 minute later

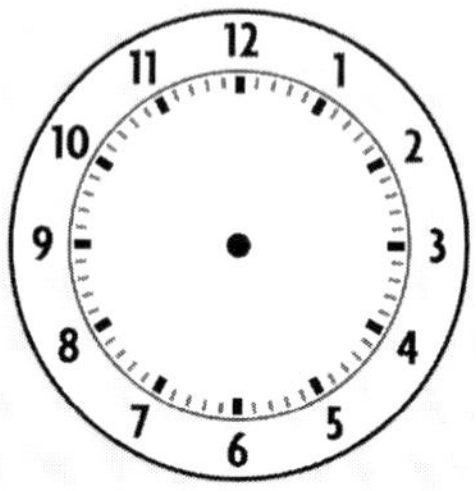

3 minutes later

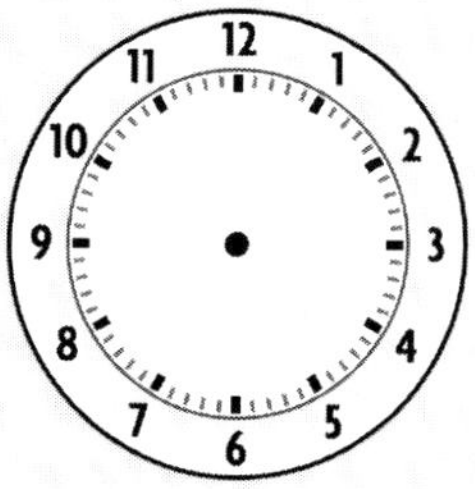

Write the time below the clock showing the hands, then write the correct past time below the empty clock and draw the hands.

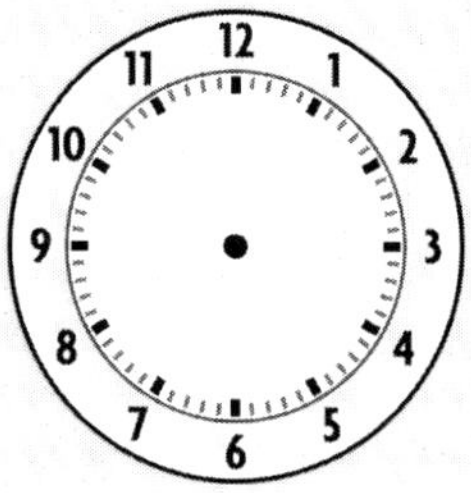

1 minute earlier

Name ______________________________ Date ____________________

Times Ending in 1-Minute Marks (:00, :01, :02, :03, ...)

Write the time below the clock showing the hands, then write the correct future time below the empty clock and draw the hands.

4 minutes later

12 minutes later

Write the time below the clock showing the hands, then write the correct past time below the empty clock and draw the hands.

3 minutes earlier

Name ______________________ Date ______________

Times Ending in 1-Minute Marks (:00, :01, :02, :03, ...)

Write the time below the clock showing the hands, then write the correct future time below the empty clock and draw the hands.

1 hour later

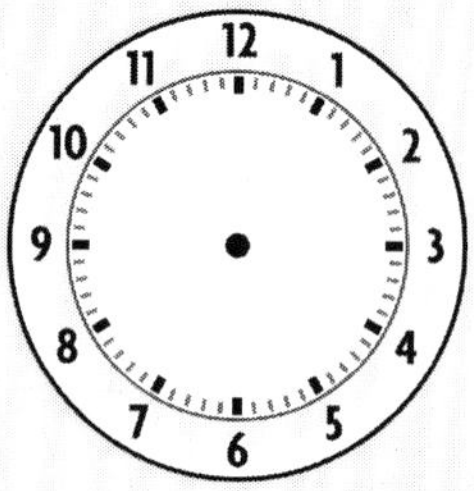

8 minutes later

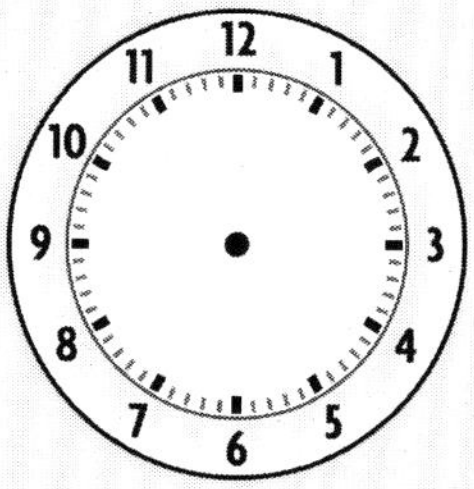

Write the time below the clock showing the hands, then write the correct past time below the empty clock and draw the hands.

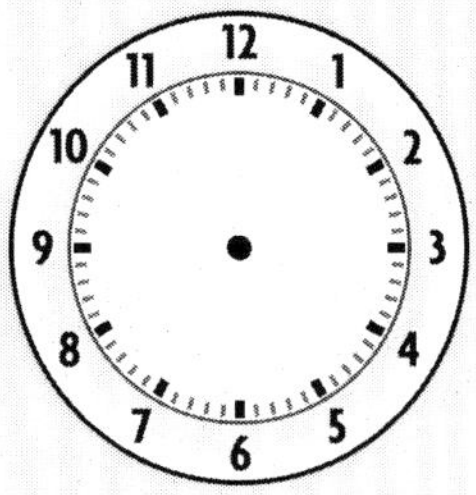

1 hour earlier

Name ______________________ Date ______________

Times Ending in 1-Minute Marks (:00, :01, :02, :03, ...)

For each event, write the beginning and ending times under the clocks, then draw the hands on the clocks to match the correct times.

1. The show started at 2:00 and ended 23 minutes later.

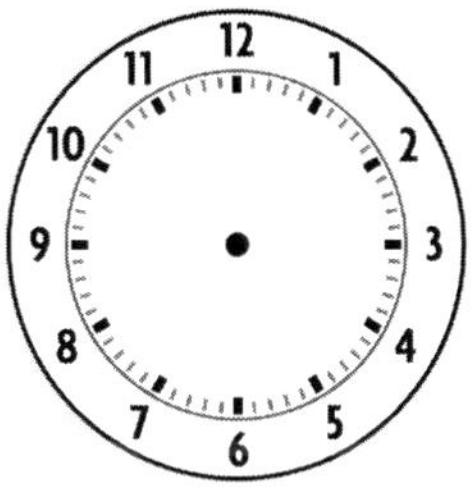
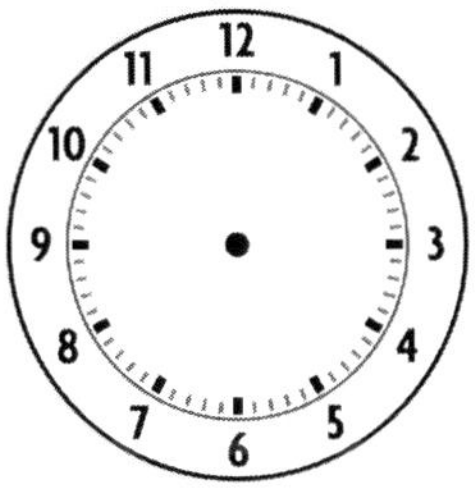

2. The friends talked on the phone for 31 minutes. They started at 5:00.

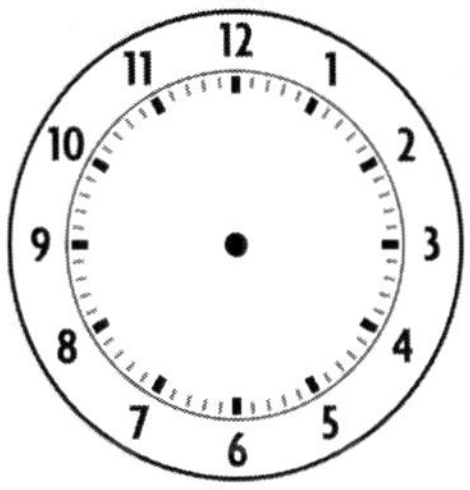
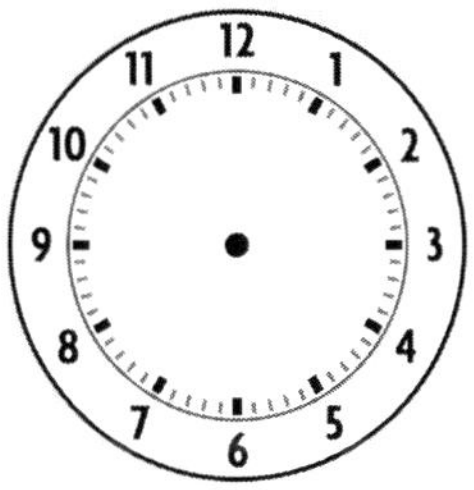

3. Jacob finished the math worksheet in 4 minutes. He started at 9:20.

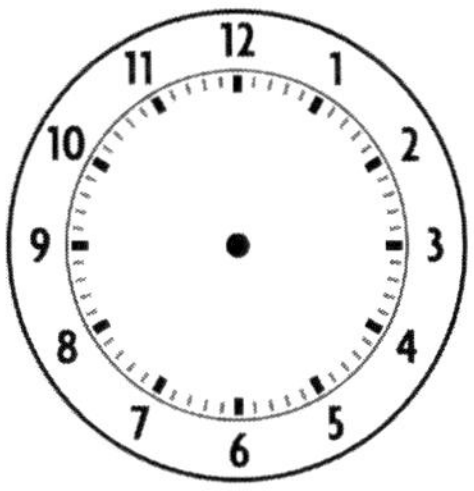
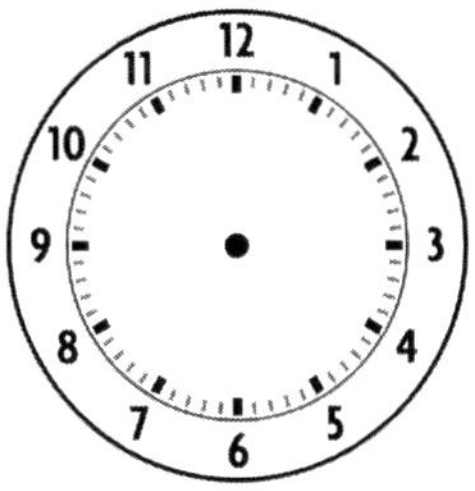

4. Jessica drank her orange juice in 3 minutes. She was finished at 7:05.

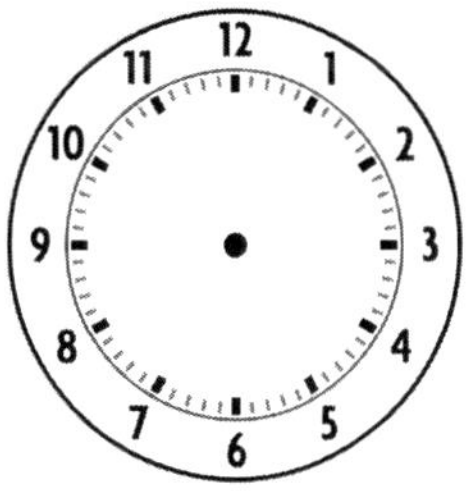
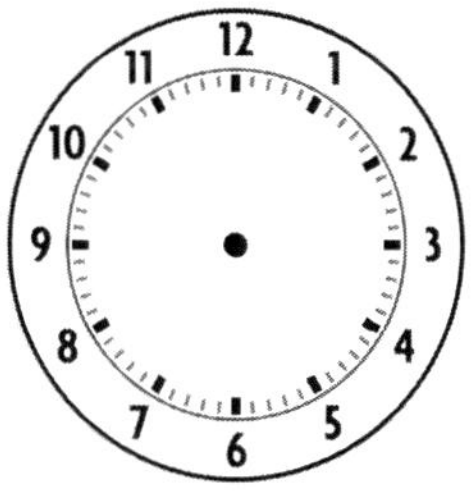

Name ______________________ Date ______________

Times Ending in 1-Minute Marks (:00, :01, :02, :03, ...)

For each event, write the beginning and ending times under the clocks, then draw the hands on the clocks to match the correct times.

1. James cleaned the table in 3 minutes. He started at 10:05.

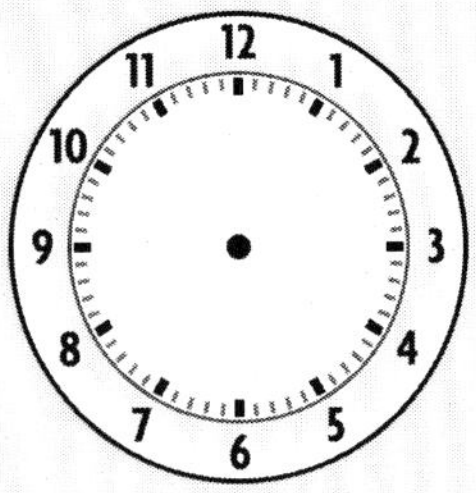

2. Rose began singing at 1:10 and finished the song in 4 minutes.

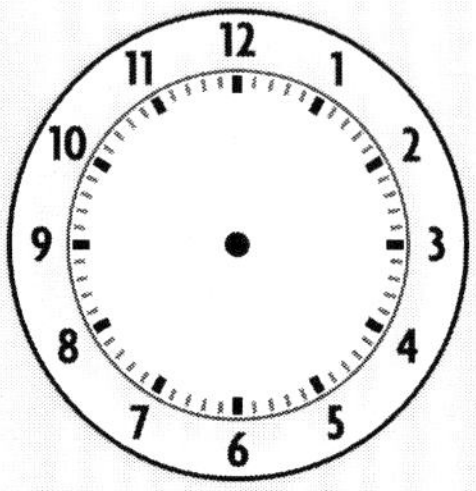

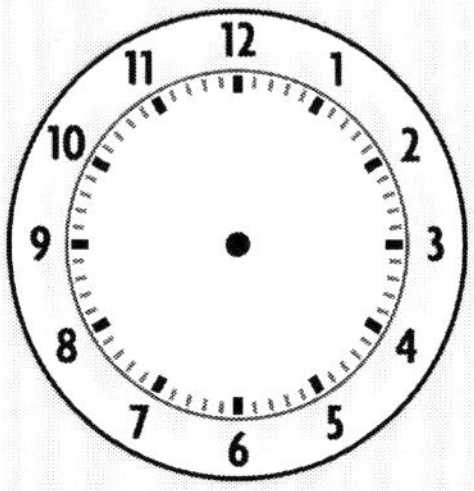

3. Luke started dancing at 5:02. He danced for 10 minutes.

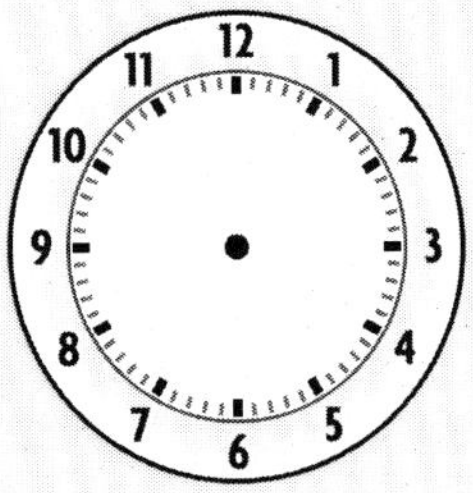

4. Mara ate a cookie in 1 minute. She was finished at 12:24.

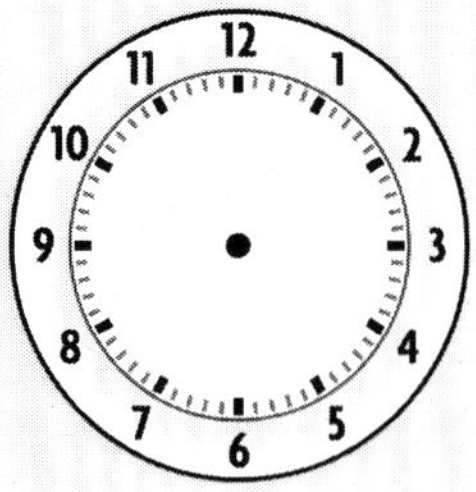

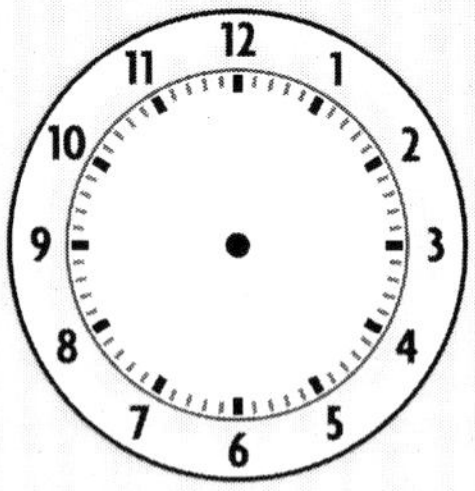

Name ______________________ Date ______________

Times Ending in 1-Minute Marks (:00, :01, :02, :03, ...)

For each event, write the beginning and ending times under the clocks, then draw the hands on the clocks to match the correct times.

1. The ant began to walk across the sidewalk at 8:22 and arrived at the other side 1 minute later.

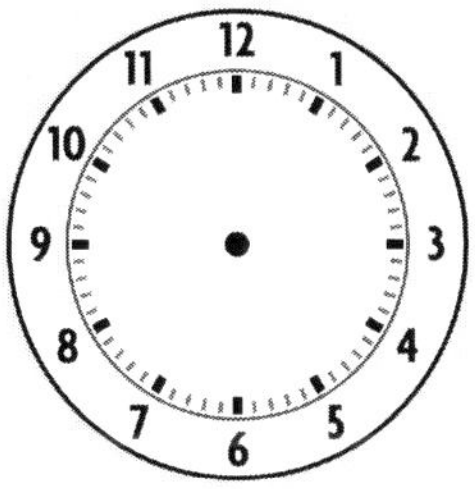

2. The cat started chasing a mouse at 3:20 and gave up 4 minutes later.

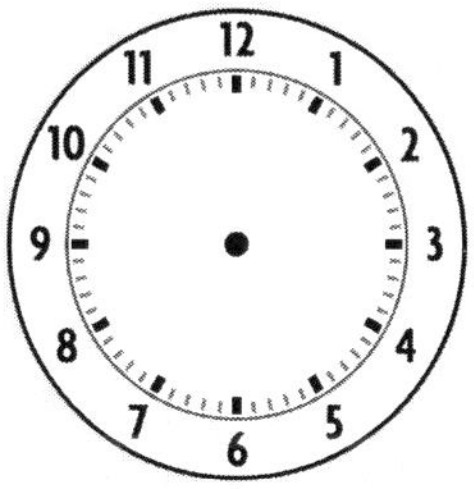

3. The chef kneaded dough for 8 minutes. He began kneading at 2:20.

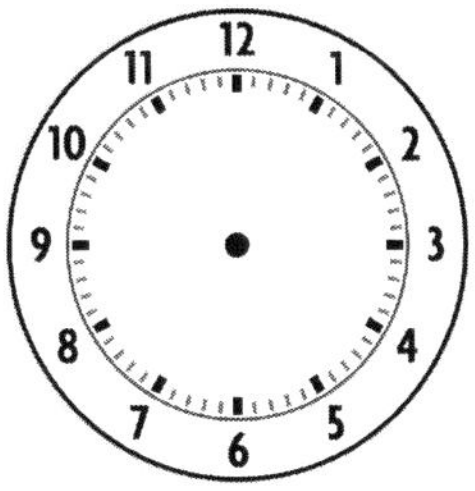

4. The students did jumping jacks for 3 minutes. They stopped at 4:05.

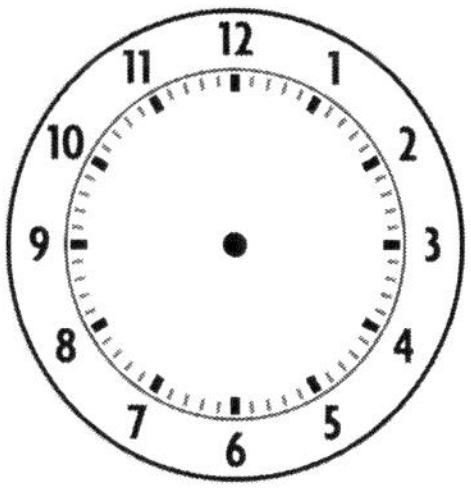

Name ________________________________ Date ____________________

Times Ending in 1-Minute Marks (:00, :01, :02, :03, ...)

For each event, write the beginning and ending times under the clocks, then draw the hands on the clocks to match the correct times.

1. The dog began to chew on a bone at 12:00 and stopped 7 minutes later.

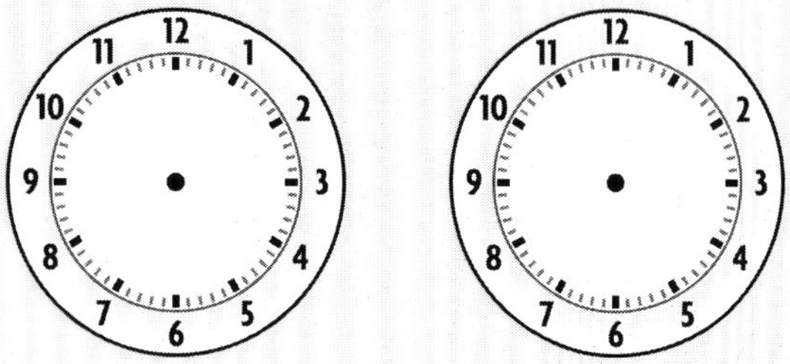

2. The students lined up to walk outside at 10:30. They were on the playground 4 minutes later.

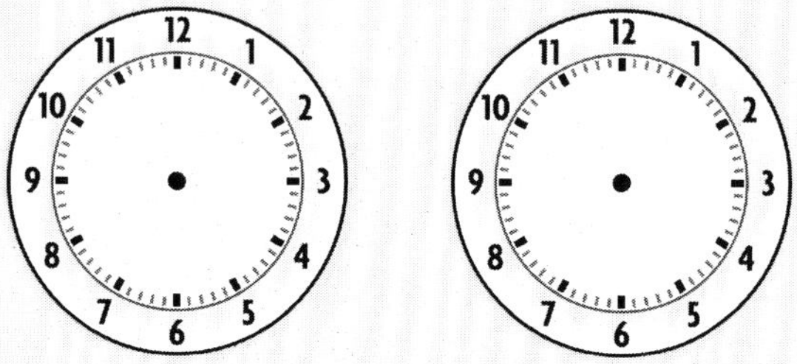

3. The children spent 2 minutes washing their hands. They started at 2:30.

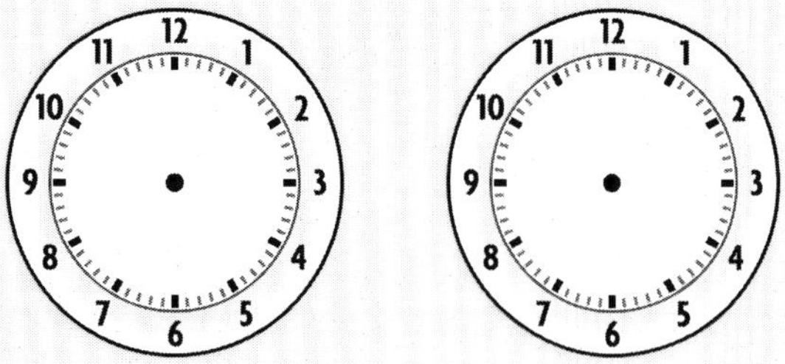

4. Daniel took 3 minutes to water the plant and stopped at 1:13.

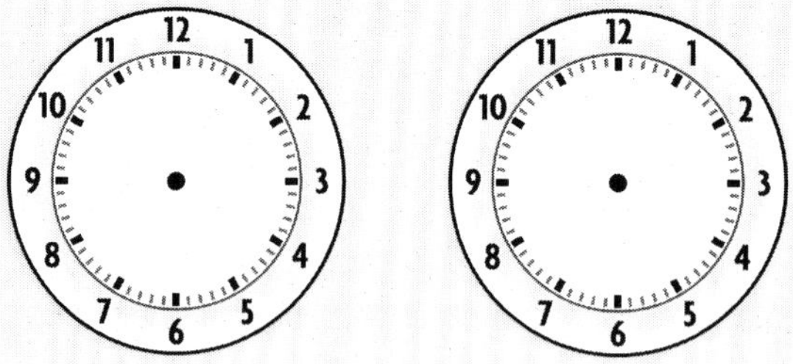

SECTION

S

SOLUTIONS

For Sections: 1-7

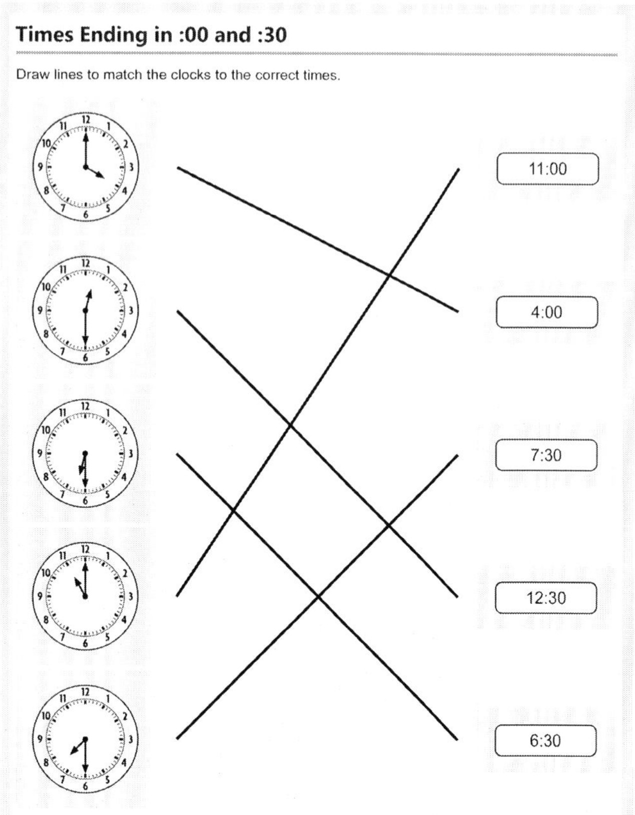
Times Ending in :00 and :30

Draw lines to match the clocks to the correct times.

11:00

4:00

7:30

12:30

6:30

PAGE 2

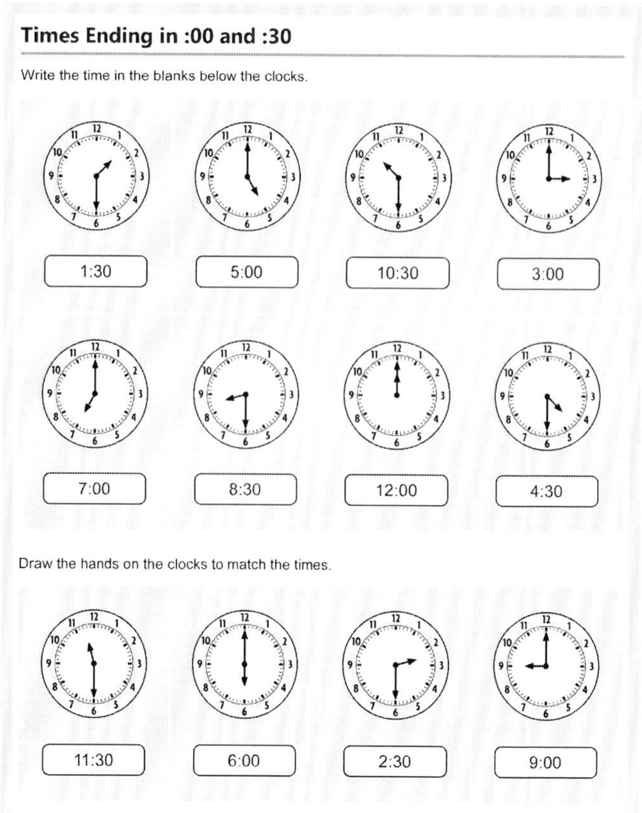
Times Ending in :00 and :30

Write the time in the blanks below the clocks.

1:30 | 5:00 | 10:30 | 3:00

7:00 | 8:30 | 12:00 | 4:30

Draw the hands on the clocks to match the times.

11:30 | 6:00 | 2:30 | 9:00

PAGE 3

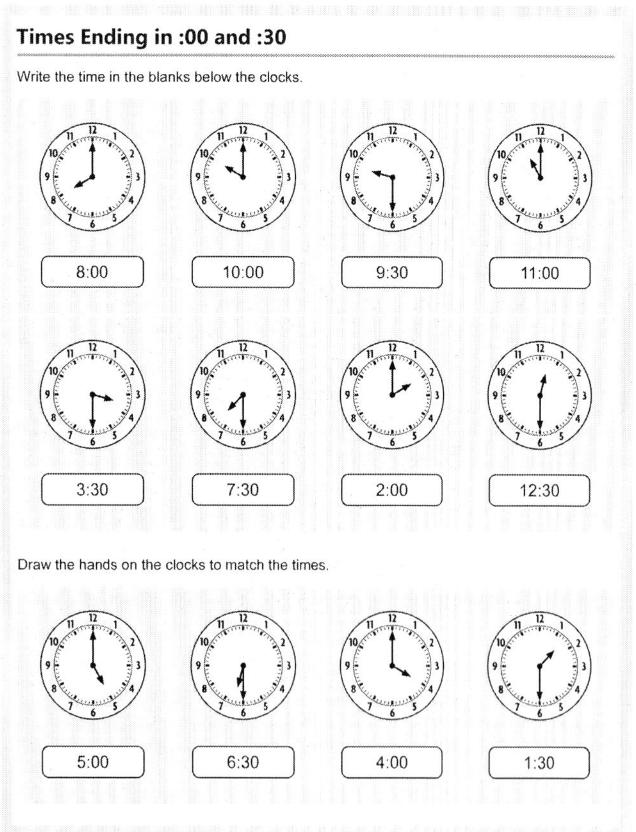
Times Ending in :00 and :30

Write the time in the blanks below the clocks.

8:00 | 10:00 | 9:30 | 11:00

3:30 | 7:30 | 2:00 | 12:30

Draw the hands on the clocks to match the times.

5:00 | 6:30 | 4:00 | 1:30

PAGE 4

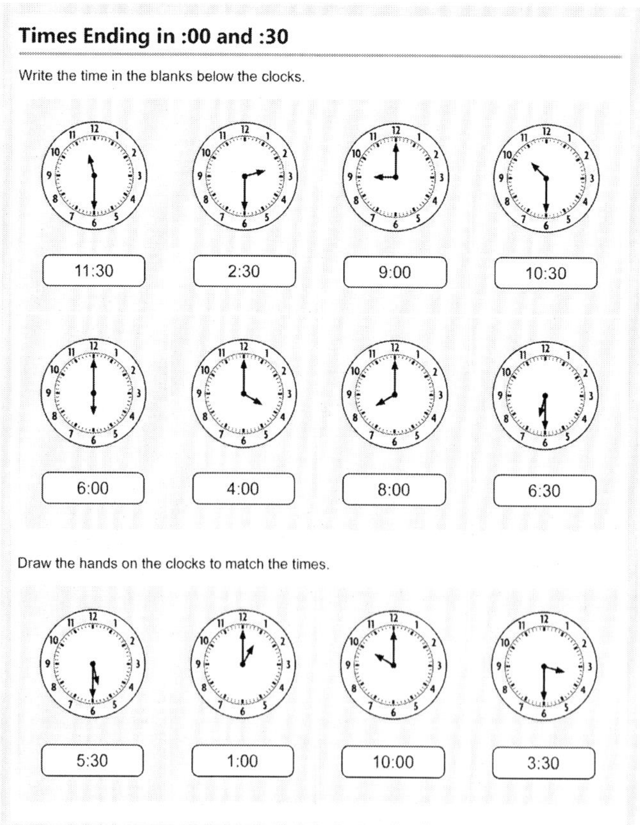
Times Ending in :00 and :30

Write the time in the blanks below the clocks.

11:30 | 2:30 | 9:00 | 10:30

6:00 | 4:00 | 8:00 | 6:30

Draw the hands on the clocks to match the times.

5:30 | 1:00 | 10:00 | 3:30

PAGE 5

Times Ending in :00 and :30

Write the time below the clock showing the hands, then write the correct future time below the empty clock and draw the hands.

30 minutes later

7:00 | 7:30

1 hour later

11:00 | 12:00

Write the time below the clock showing the hands, then write the correct past time below the empty clock and draw the hands.

1 hour earlier

2:00 | 3:00

PAGE 6

Times Ending in :00 and :30

Write the time below the clock showing the hands, then write the correct future time below the empty clock and draw the hands.

2 hours later

6:00 | 8:00

30 minutes later

2:00 | 2:30

Write the time below the clock showing the hands, then write the correct past time below the empty clock and draw the hands.

30 minutes earlier

4:00 | 4:30

PAGE 7

Times Ending in :00 and :30

For each event, write the beginning and ending times under the clocks, then draw the hands on the clocks to match the correct times.

1. Emma wakes up at 6:00 and arrives at school 2 hours later.

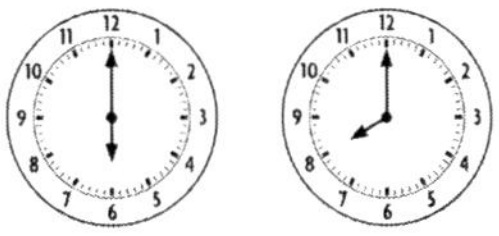

6:00 | 8:00

2. Greg's soccer practice starts at 4:00 and lasts 1 hour.

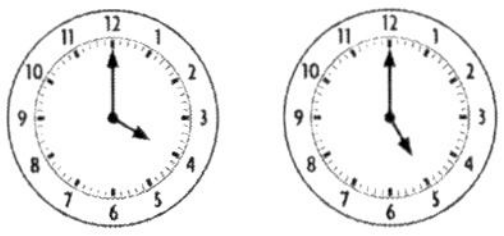

4:00 | 5:00

3. Lunch begins at 12:00 and lasts 30 minutes.

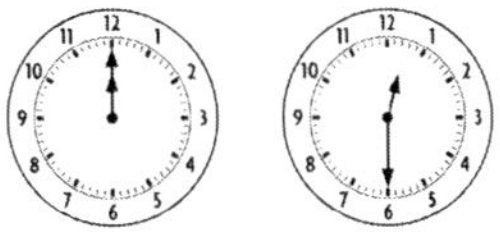

12:00 | 12:30

4. James played video games for 1 hour and finished playing at 10:00.

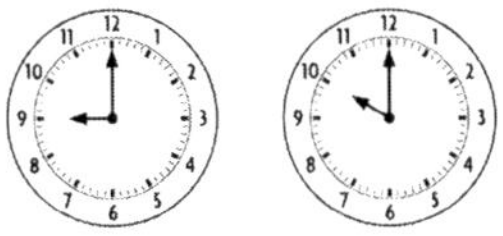

9:00 | 10:00

PAGE 8

Times Ending in :15

Write the time in the blanks below the clocks.

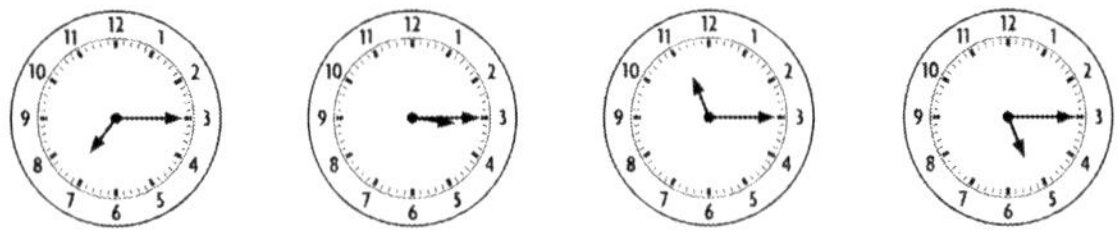

7:15 | 3:15 | 11:15 | 5:15

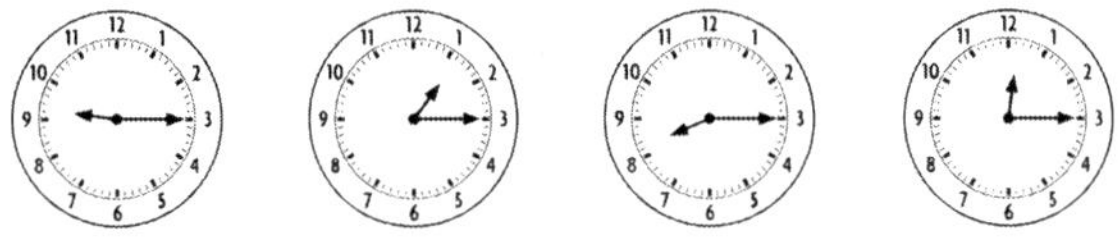

9:15 | 1:15 | 8:15 | 12:15

Draw the hands on the clocks to match the times.

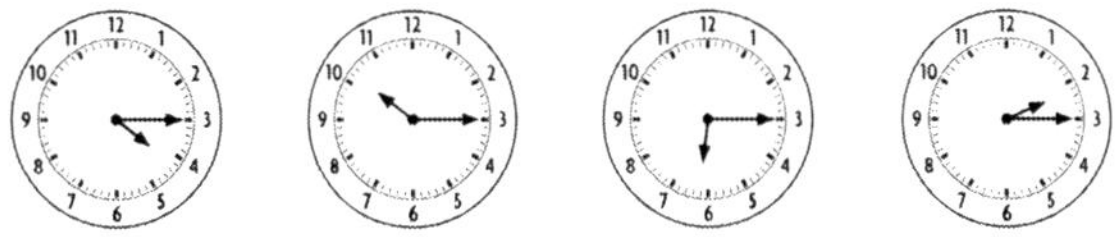

4:15 | 10:15 | 6:15 | 2:15

PAGE 10

Times Ending in :15

Write the time below the clock showing the hands, then write the correct future time below the empty clock and draw the hands.

Write the time below the clock showing the hands, then write the correct past time below the empty clock and draw the hands.

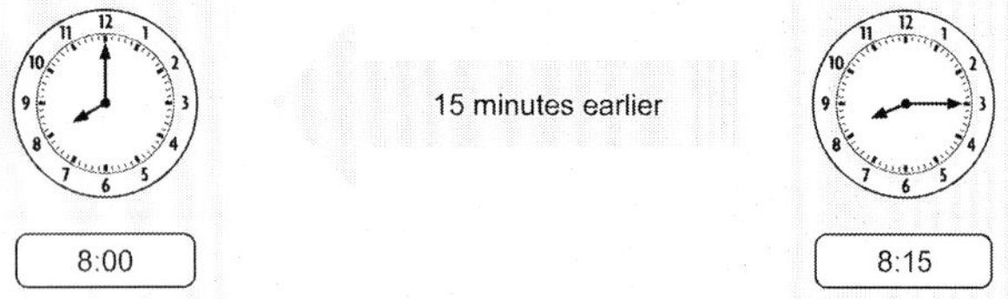

PAGE 11

Times Ending in :15

For each event, write the beginning and ending times under the clocks, then draw the hands on the clocks to match the correct times.

1. John leaves at 7:00 and gets to school 15 minutes later.

2. It is 3:00 and Julie's art class begins in 15 minutes.

3. Basketball practice begins at 4:15 and lasts 1 hour.

4. Olivia jogs for 15 minutes. She finishes at 10:15.

PAGE 12

Times Ending in :00, :15, :30

Draw lines to match the clocks to the correct times.

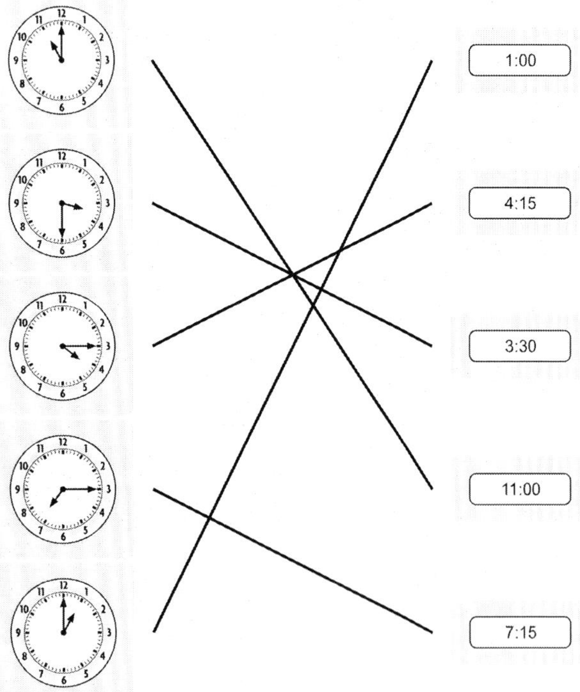

PAGE 14

Times Ending in :00, :15, :30

Draw lines to match the clocks to the correct times.

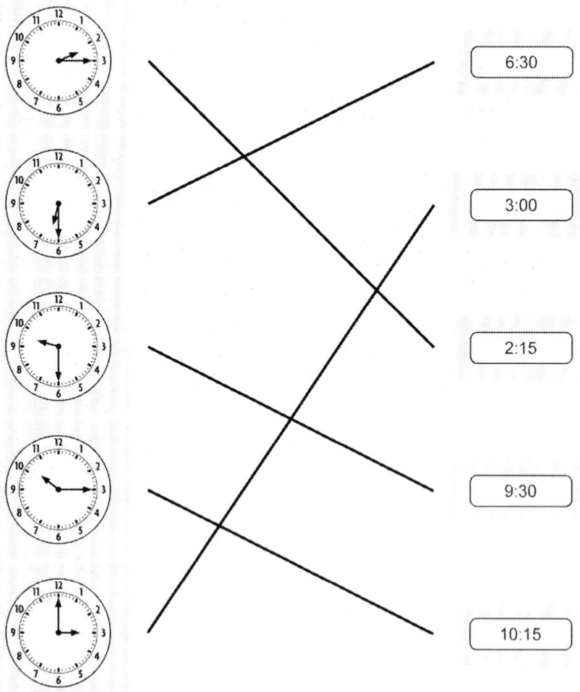

PAGE 15

Times Ending in :00, :15, :30

Write the time in the blanks below the clocks.

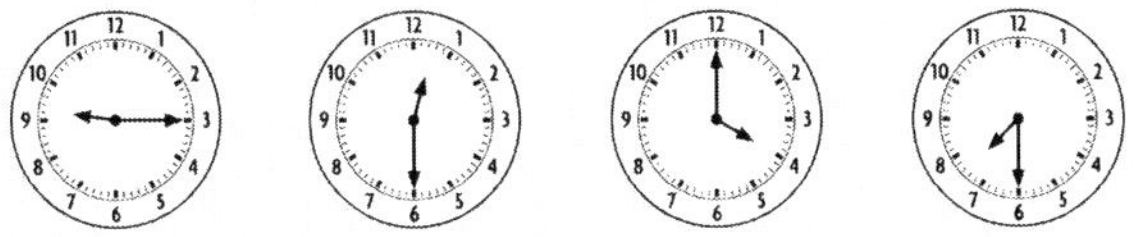

9:15	12:30	4:00	7:30

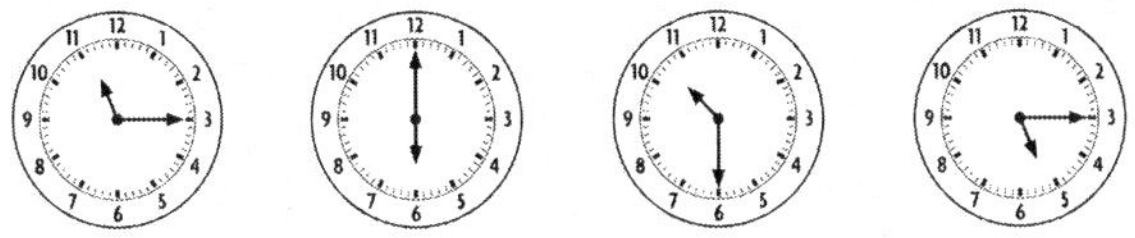

11:15	6:00	10:30	5:15

Draw the hands on the clocks to match the times.

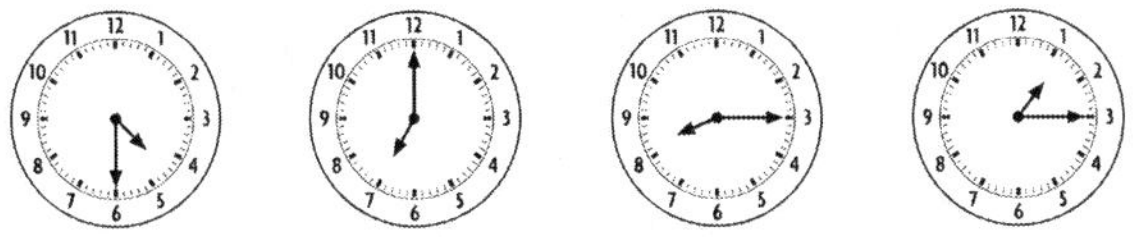

4:30	7:00	8:15	1:15

PAGE 16

Times Ending in :00, :15, :30

Write the time in the blanks below the clocks.

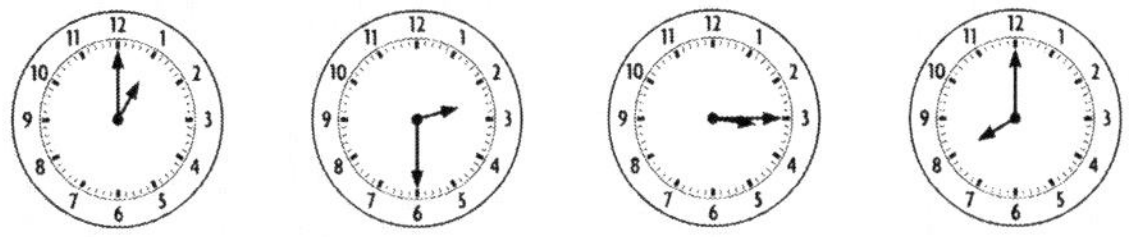

1:00	2:30	3:15	8:00

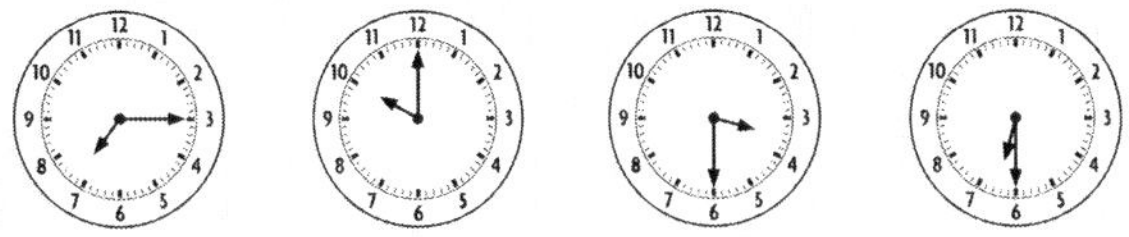

7:15	10:00	3:30	6:30

Draw the hands on the clocks to match the times.

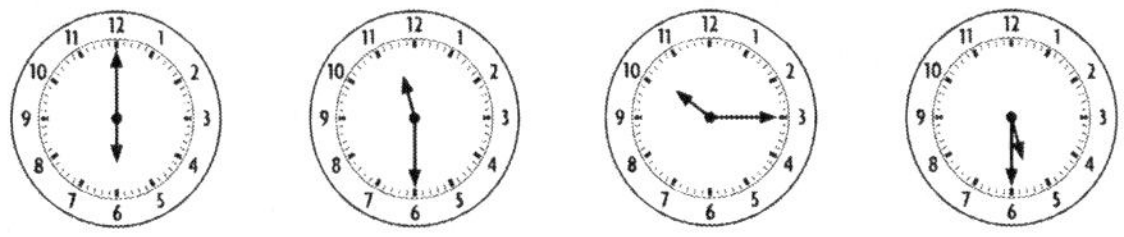

6:00	11:30	10:15	5:30

PAGE 17

Times Ending in :00, :15, :30

Write the time in the blanks below the clocks.

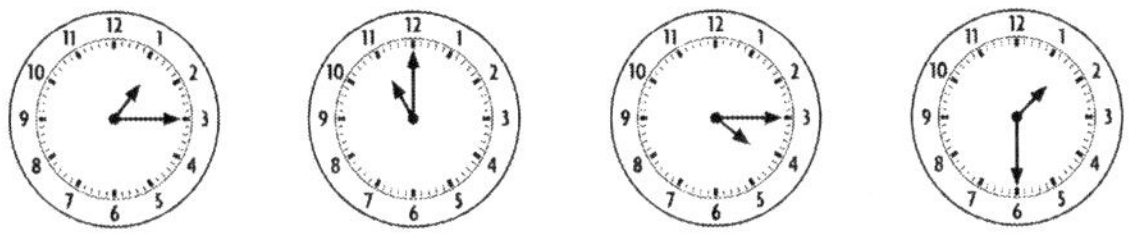

1:15	11:00	4:15	1:30

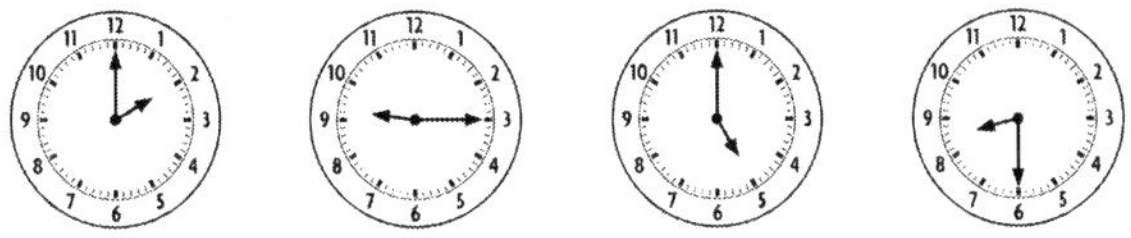

2:00	9:15	5:00	8:30

Draw the hands on the clocks to match the times.

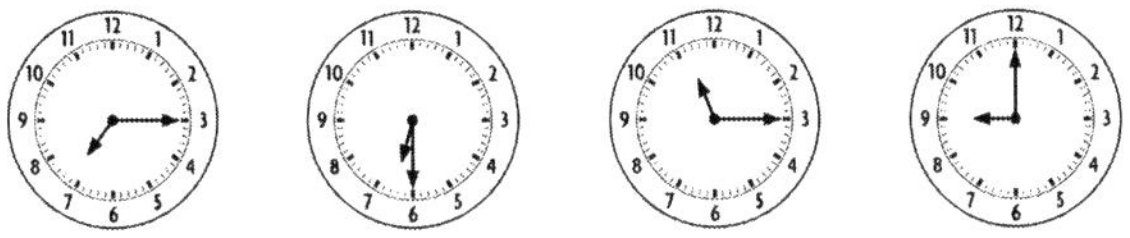

7:15	6:30	11:15	9:00

PAGE 18

Times Ending in :00, :15, :30

Write the time in the blanks below the clocks.

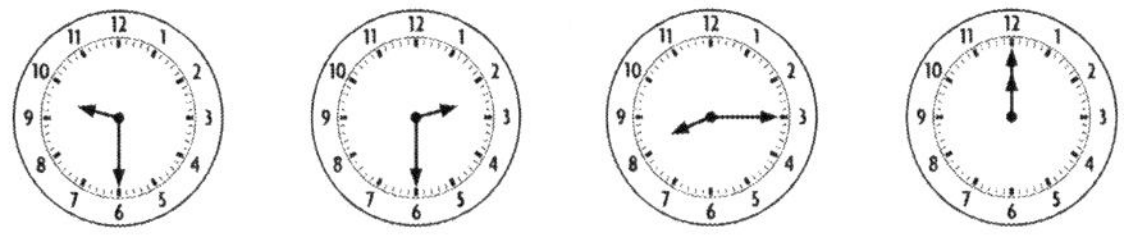

9:30	2:30	8:15	12:00

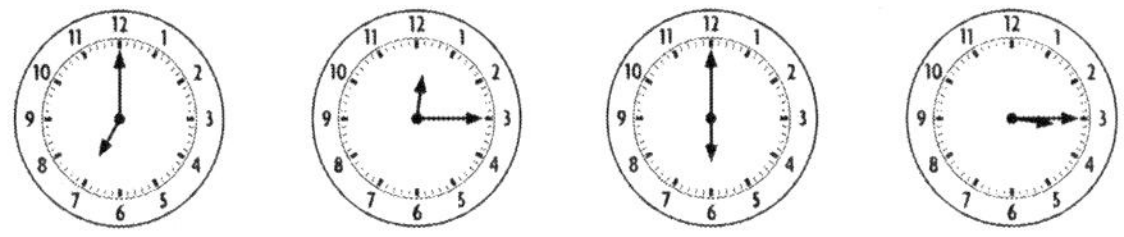

7:00	12:15	6:00	3:15

Draw the hands on the clocks to match the times.

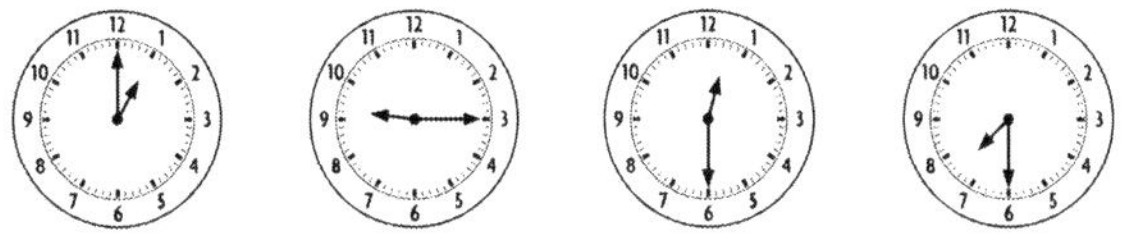

1:00	9:15	12:30	7:30

PAGE 19

Times Ending in :00, :15, :30

Write the time in the blanks below the clocks.

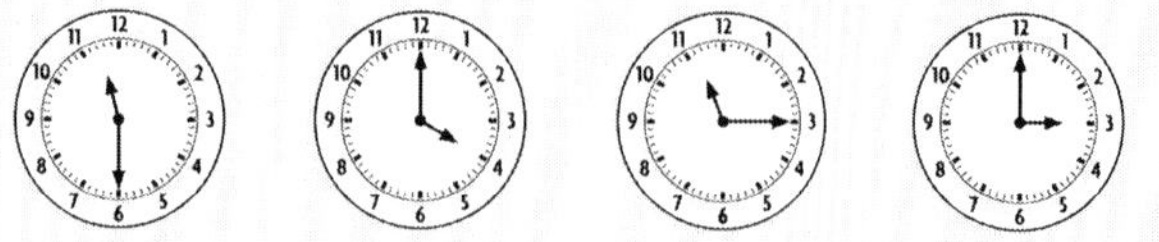

11:30 | 4:00 | 11:15 | 3:00

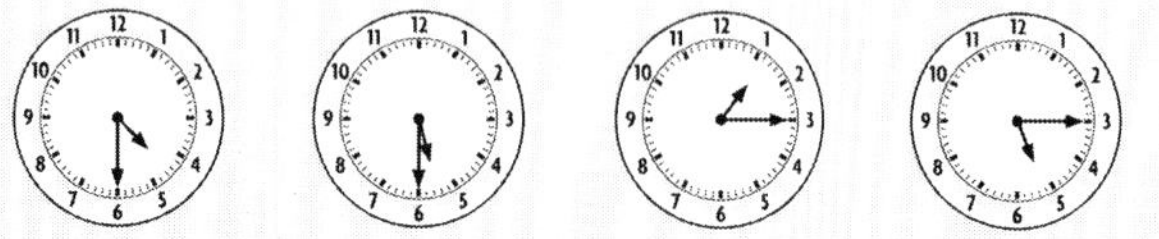

4:30 | 5:30 | 1:15 | 5:15

Draw the hands on the clocks to match the times.

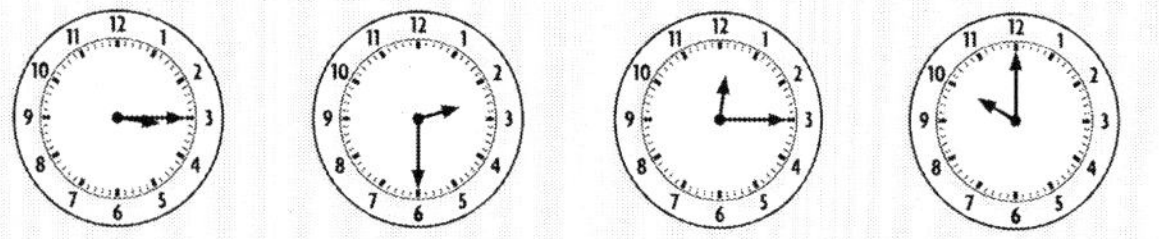

3:15 | 2:30 | 12:15 | 10:00

PAGE 20

Times Ending in :00, :15, :30

Write the time in the blanks below the clocks.

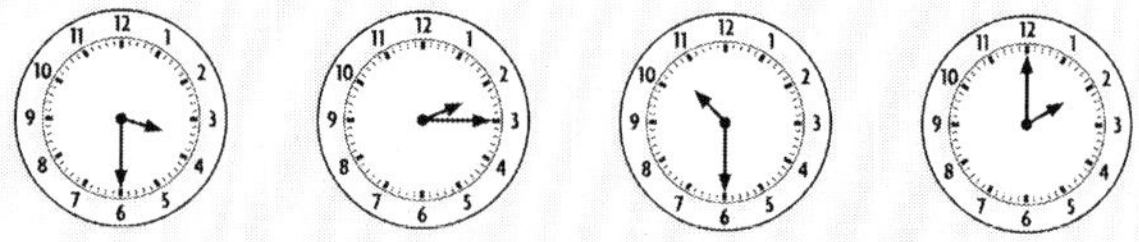

3:30 | 2:15 | 10:30 | 2:00

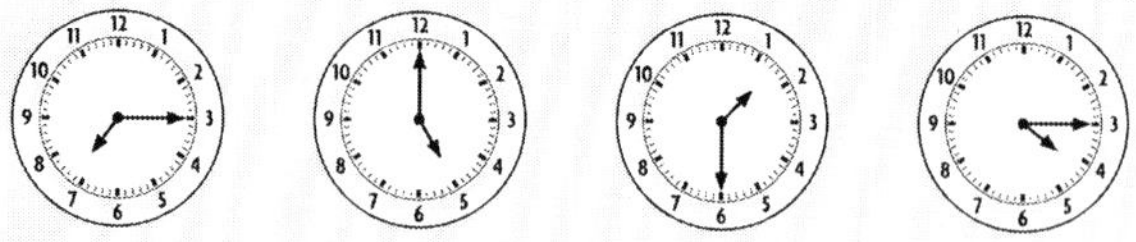

7:15 | 5:00 | 1:30 | 4:15

Draw the hands on the clocks to match the times.

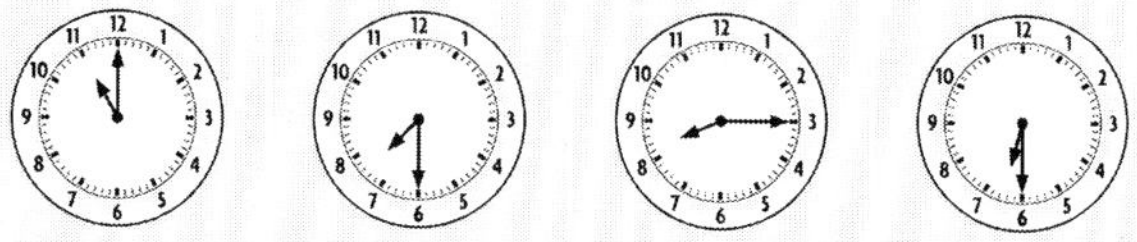

11:00 | 7:30 | 8:15 | 6:30

PAGE 21

Times Ending in :00, :15, :30

Write the time in the blanks below the clocks.

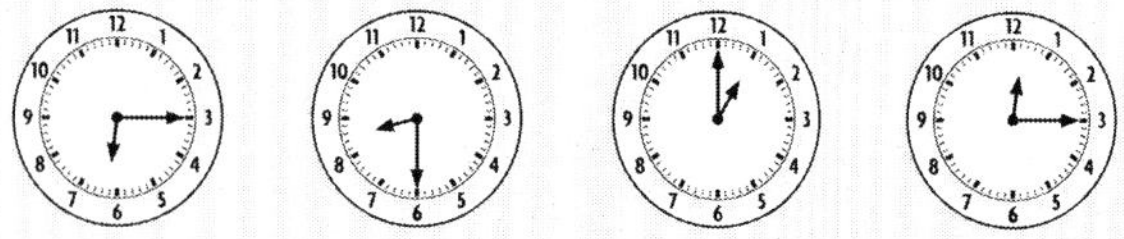

6:15 | 8:30 | 1:00 | 12:15

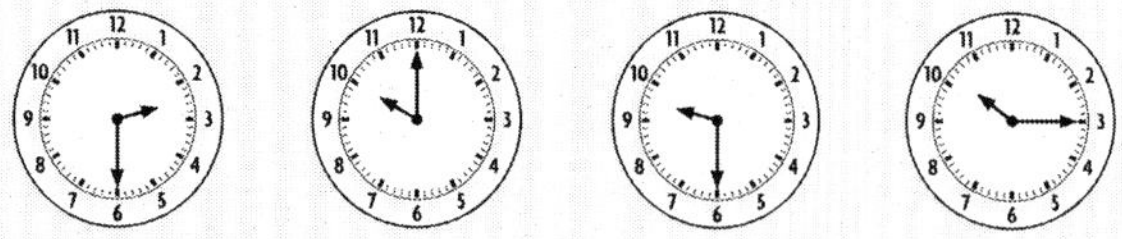

2:30 | 10:00 | 9:30 | 10:15

Draw the hands on the clocks to match the times.

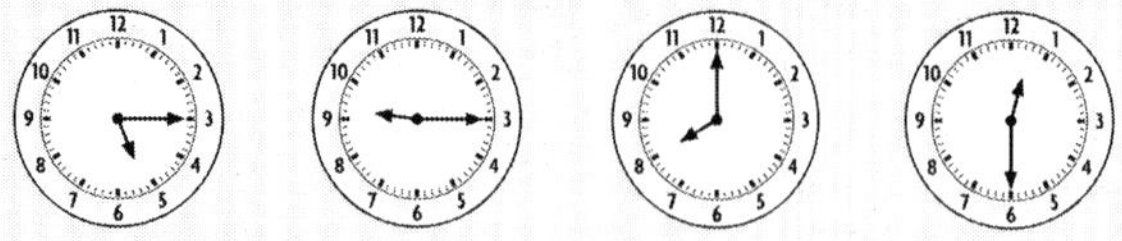

5:15 | 9:15 | 8:00 | 12:30

PAGE 22

Times Ending in :00, :15, :30

Write the time below the clock showing the hands, then write the correct future time below the empty clock and draw the hands.

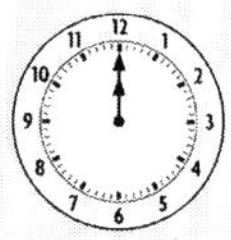

12:00

30 minutes later

12:30

1:15

1 hour later

2:15

Write the time below the clock showing the hands, then write the correct past time below the empty clock and draw the hands.

4:00

30 minutes earlier

4:30

PAGE 23

Times Ending in :00, :15, :30

Write the time below the clock showing the hands, then write the correct future time below the empty clock and draw the hands.

2 hours later

1:00 | 3:00

30 minutes later

7:00 | 7:30

Write the time below the clock showing the hands, then write the correct past time below the empty clock and draw the hands.

15 minutes earlier

8:00 | 8:15

PAGE 24

Times Ending in :00, :15, :30

Write the time below the clock showing the hands, then write the correct future time below the empty clock and draw the hands.

1 hour later

10:30 | 11:30

2 hours later

2:00 | 4:00

Write the time below the clock showing the hands, then write the correct past time below the empty clock and draw the hands.

15 minutes earlier

5:00 | 5:15

PAGE 25

Times Ending in :00, :15, :30

For each event, write the beginning and ending times under the clocks, then draw the hands on the clocks to match the correct times.

1. The plane took off at 2:30 and landed 1 hour later.

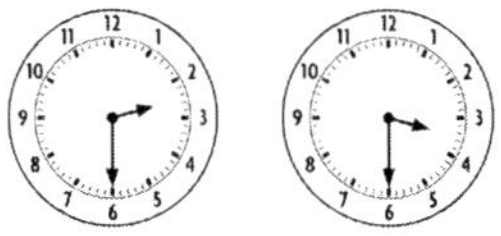

2:30 | 3:30

2. Alexis started eating at 3:00 and finished in 15 minutes.

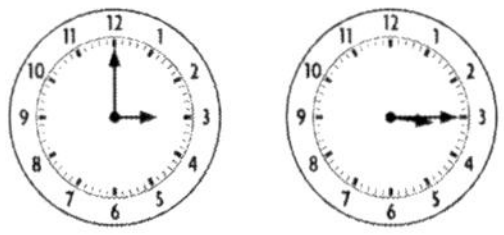

3:00 | 3:15

3. The party started at 4:15 and ended 2 hours later.

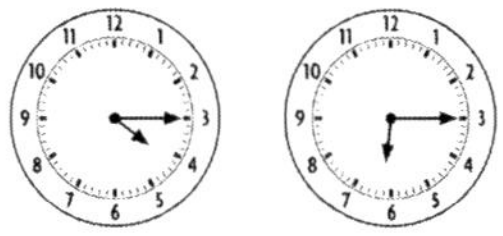

4:15 | 6:15

4. The children slept for 30 minutes and woke up at 5:30.

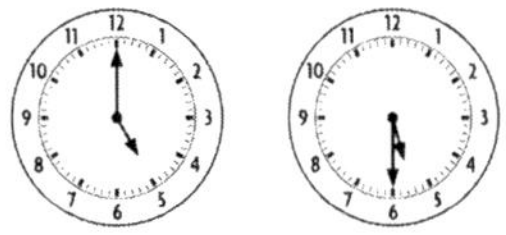

5:00 | 5:30

PAGE 26

Times Ending in :00, :15, :30

For each event, write the beginning and ending times under the clocks, then draw the hands on the clocks to match the correct times.

1. Ryan started his math homework at 4:00 and finished in 15 minutes.

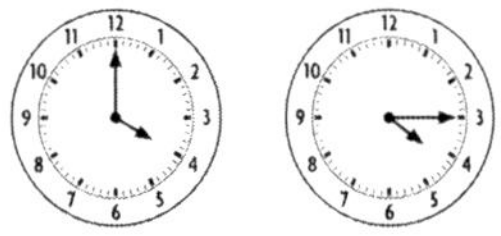

4:00 | 4:15

2. Melissa left her house at 2:00 and walked for 30 minutes.

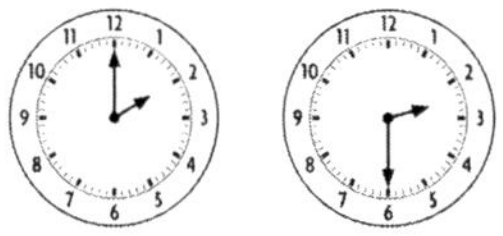

2:00 | 2:30

3. Jada's dance class begins at 6:15 and lasts 1 hour.

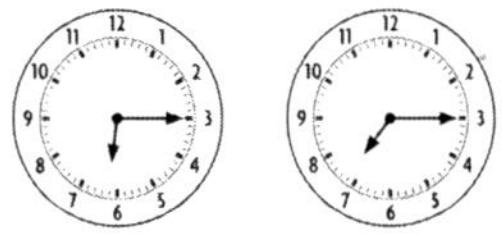

6:15 | 7:15

4. The band played for 30 minutes. They finished at 3:30.

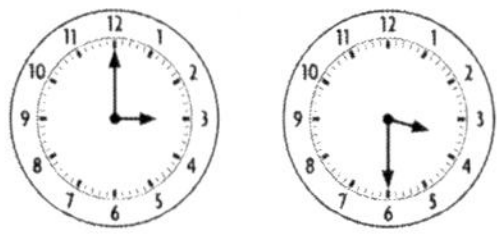

3:00 | 3:30

PAGE 27

Times Ending in :00, :15, :30

For each event, write the beginning and ending times under the clocks, then draw the hands on the clocks to match the correct times.

1. Brianna left work at 5:00 and arrived home 15 minutes later.

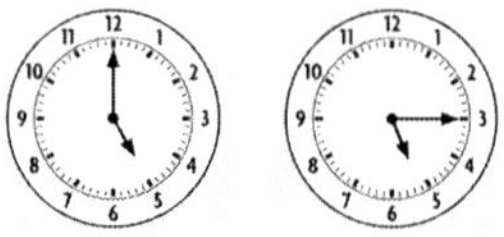

5:00 | 5:15

2. Angele began her homework at 2:30 and finished 1 hour later.

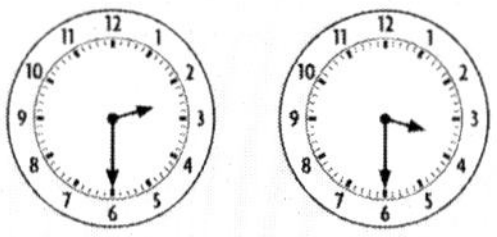

2:30 | 3:30

3. Luke started painting at 12:00 and finished in 15 minutes.

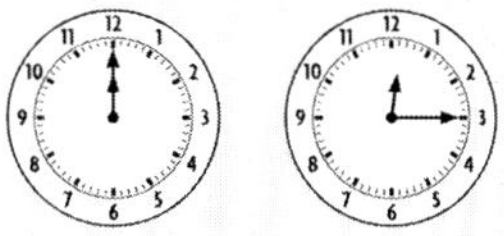

12:00 | 12:15

4. Ronny played a game for 15 minutes. He stopped playing at 6:15.

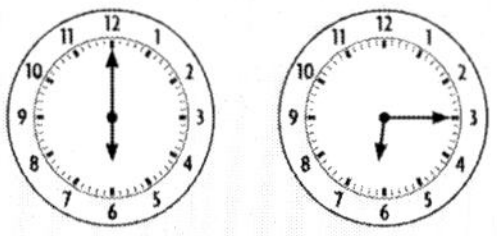

6:00 | 6:15

PAGE 28

Times Ending in :45

Write the time in the blanks below the clocks.

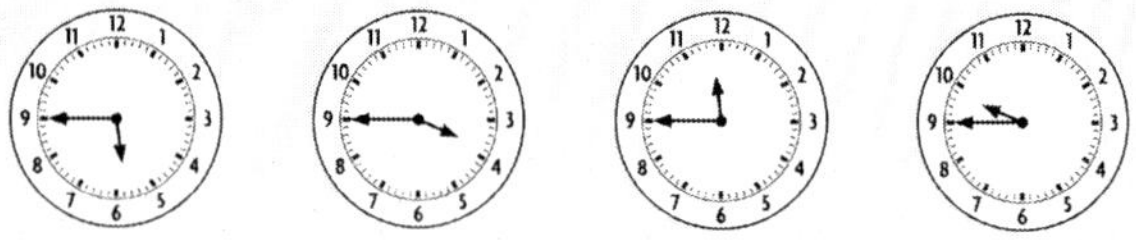

5:45 | 3:45 | 11:45 | 9:45

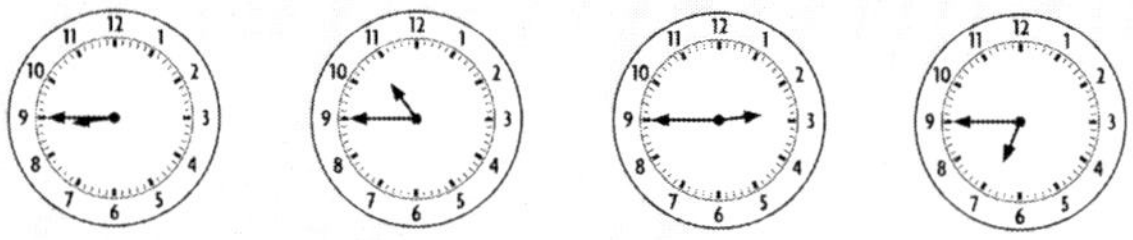

8:45 | 10:45 | 2:45 | 6:45

Draw the hands on the clocks to match the times.

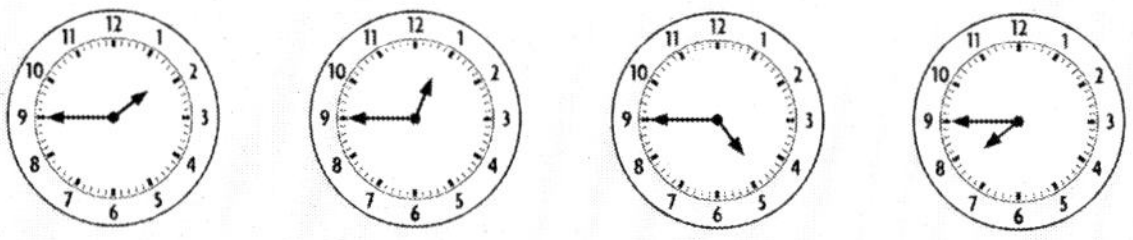

1:45 | 12:45 | 4:45 | 7:45

PAGE 30

Times Ending in :45

Write the time below the clock showing the hands, then write the correct future time below the empty clock and draw the hands.

3:00

45 minutes later

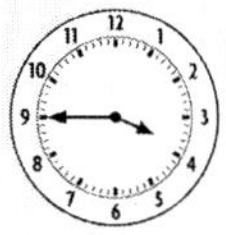

3:45

11:00

45 minutes later

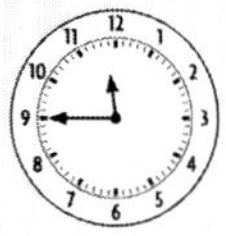

11:45

Write the time below the clock showing the hands, then write the correct past time below the empty clock and draw the hands.

4:00

45 minutes earlier

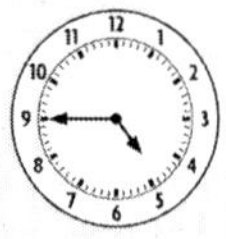

4:45

PAGE 31

Times Ending in :45

For each event, write the beginning and ending times under the clocks, then draw the hands on the clocks to match the correct times.

1. Trey starts drawing at 5:00 and he draws for 45 minutes.

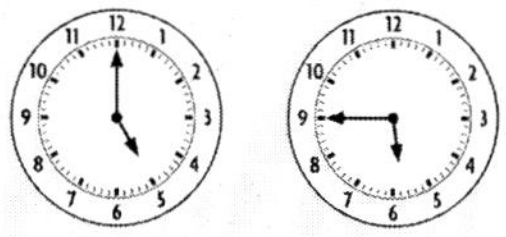

5:00 | 5:45

2. The bus left at 7:00 and stopped at Jason's house 45 minutes later.

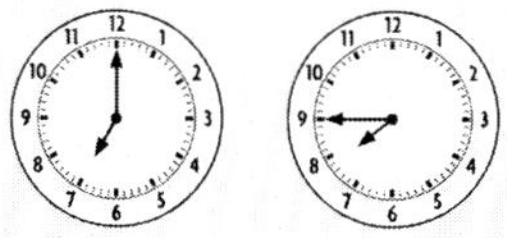

7:00 | 7:45

3. Lydia's cat took a nap at 1:45 and woke up 1 hour later.

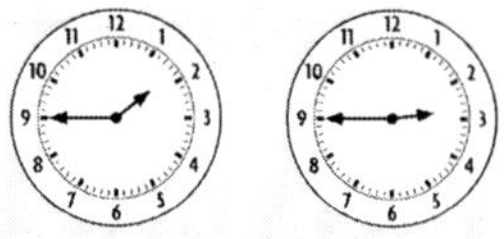

1:45 | 2:45

4. The play was 45 minutes long and ended at 6:45.

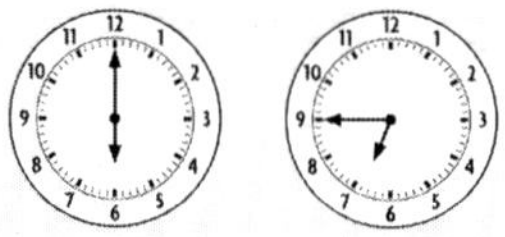

6:00 | 6:45

PAGE 32

Times Ending in :00, :15, :30, :45

Draw lines to match the clocks to the correct times.

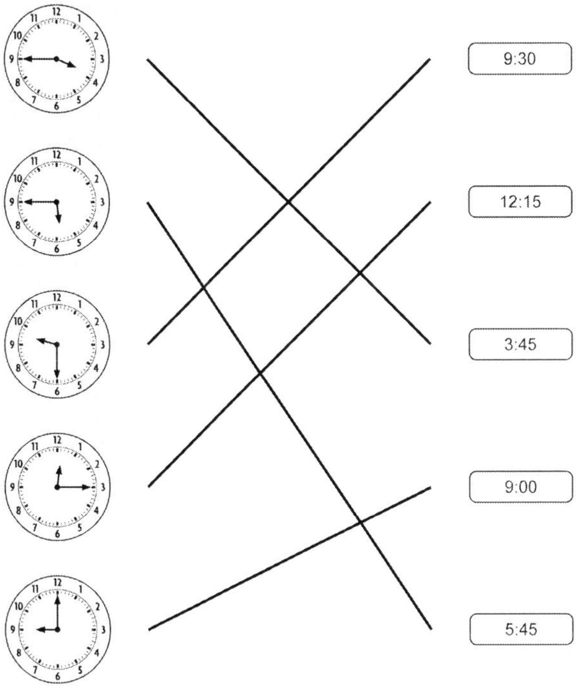

PAGE 34

Times Ending in :00, :15, :30, :45

Draw lines to match the clocks to the correct times.

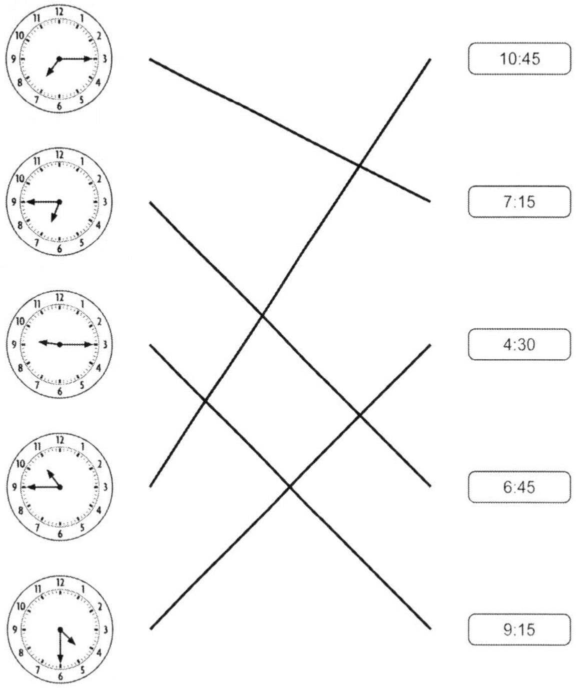

PAGE 35

Times Ending in :00, :15, :30, :45

Write the time in the blanks below the clocks.

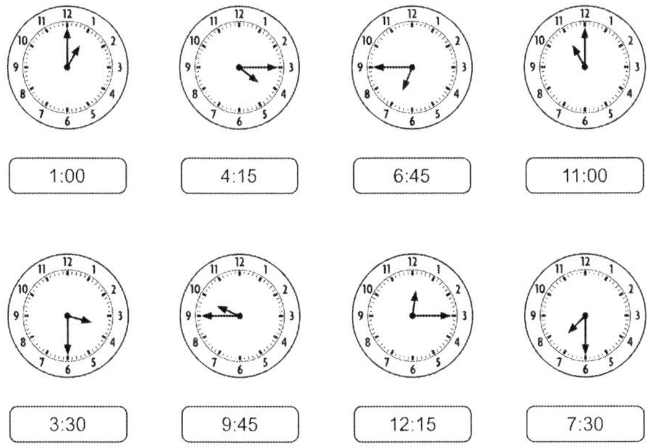

Draw the hands on the clocks to match the times.

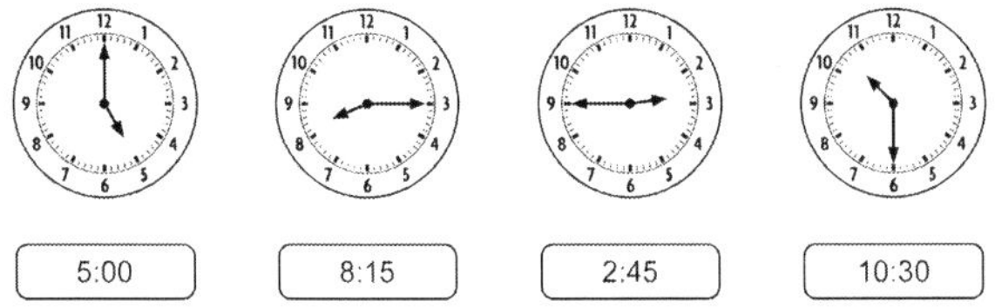

PAGE 36

Times Ending in :00, :15, :30, :45

Write the time in the blanks below the clocks.

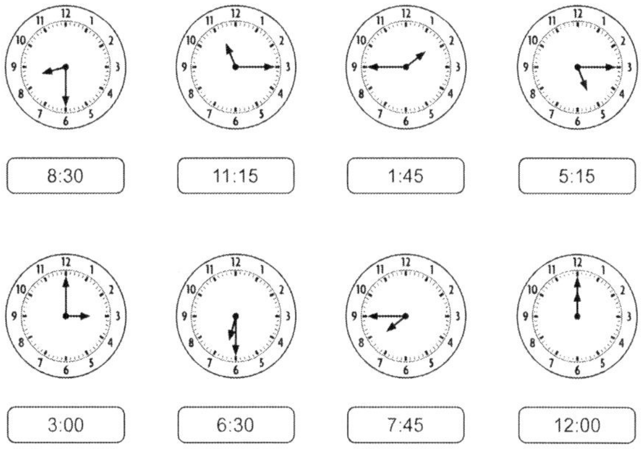

Draw the hands on the clocks to match the times.

PAGE 37

Times Ending in :00, :15, :30, :45

Write the time in the blanks below the clocks.

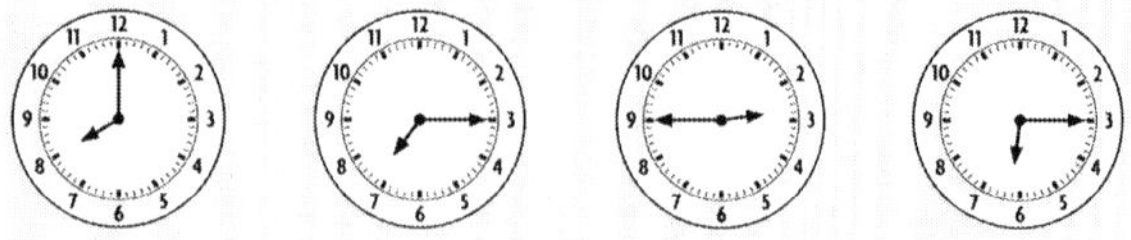

8:00 | 7:15 | 2:45 | 6:15

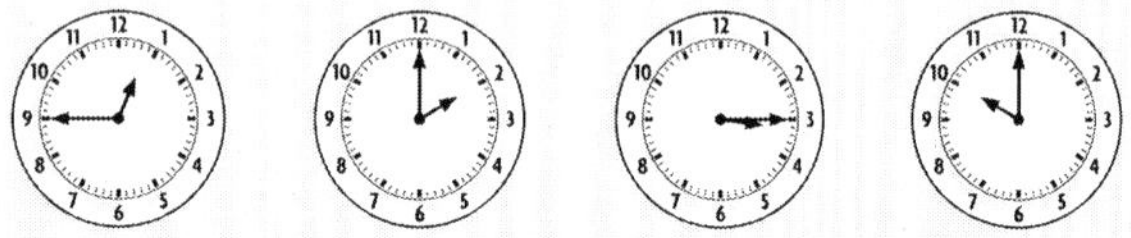

12:45 | 2:00 | 3:15 | 10:00

Draw the hands on the clocks to match the times.

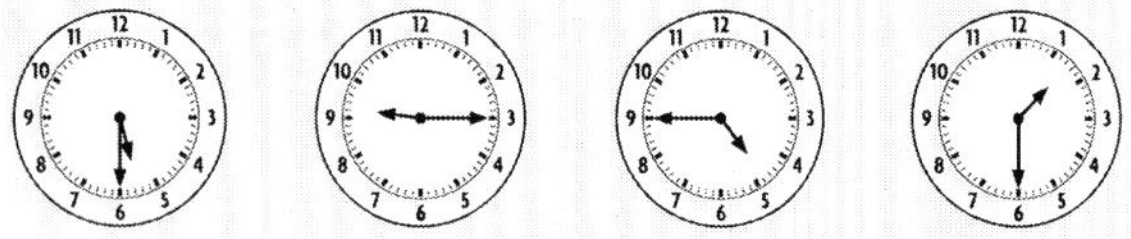

5:30 | 9:15 | 4:45 | 1:30

PAGE 38

Times Ending in :00, :15, :30, :45

Write the time in the blanks below the clocks.

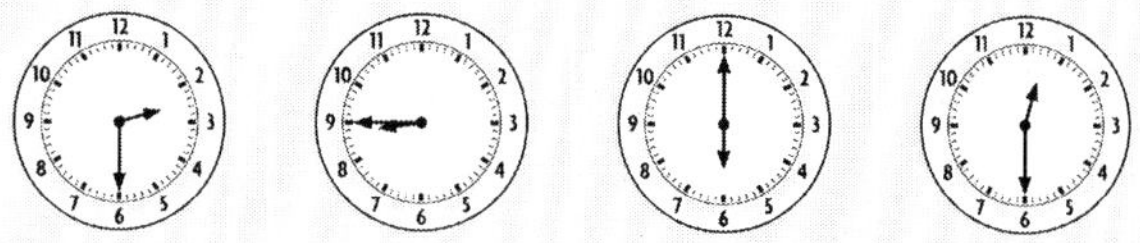

2:30 | 8:45 | 6:00 | 12:30

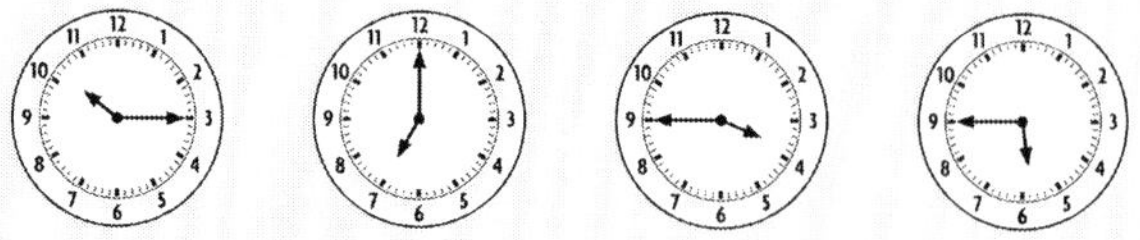

10:15 | 7:00 | 3:45 | 5:45

Draw the hands on the clocks to match the times.

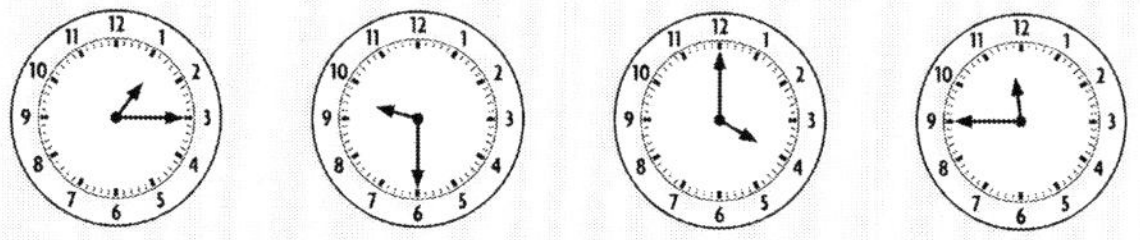

1:15 | 9:30 | 4:00 | 11:45

PAGE 39

Times Ending in :00, :15, :30, :45

Write the time in the blanks below the clocks.

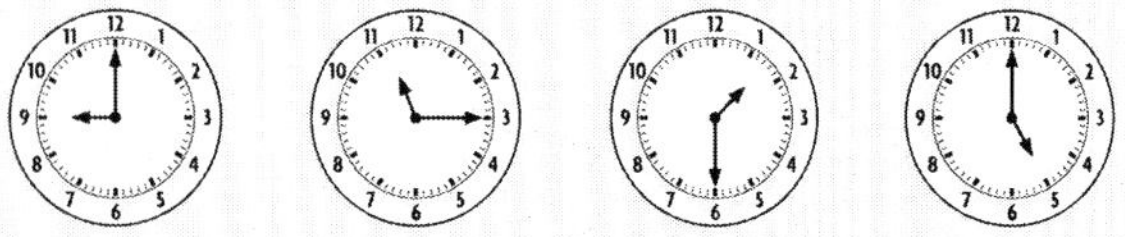

9:00 | 11:15 | 1:30 | 5:00

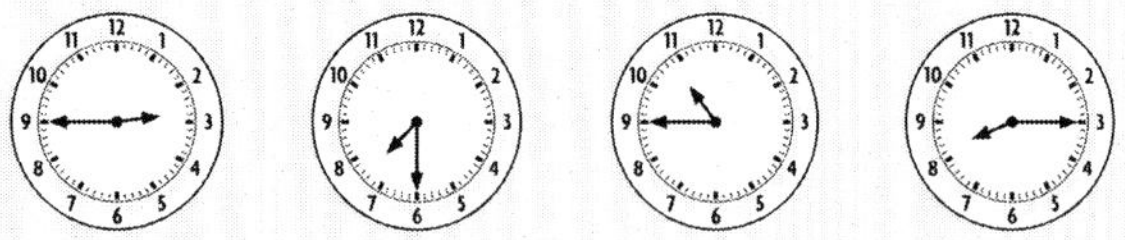

2:45 | 7:30 | 10:45 | 8:15

Draw the hands on the clocks to match the times.

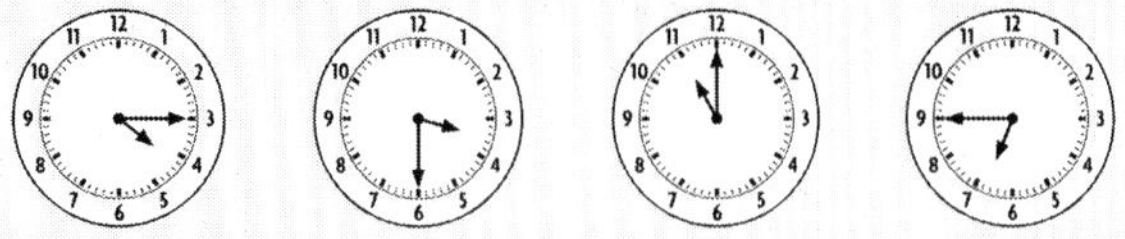

4:15 | 3:30 | 11:00 | 6:45

PAGE 40

Times Ending in :00, :15, :30, :45

Write the time in the blanks below the clocks.

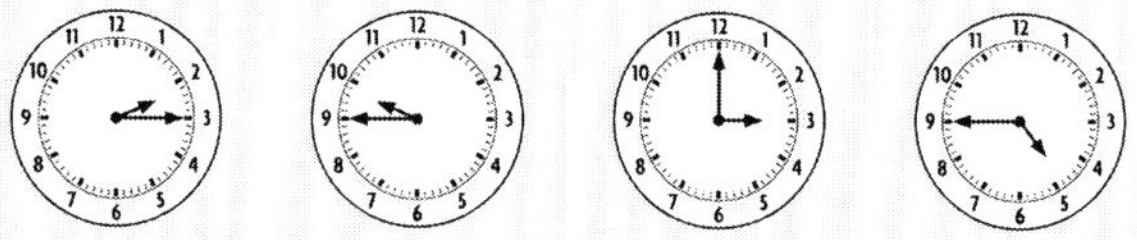

2:15 | 9:45 | 3:00 | 4:45

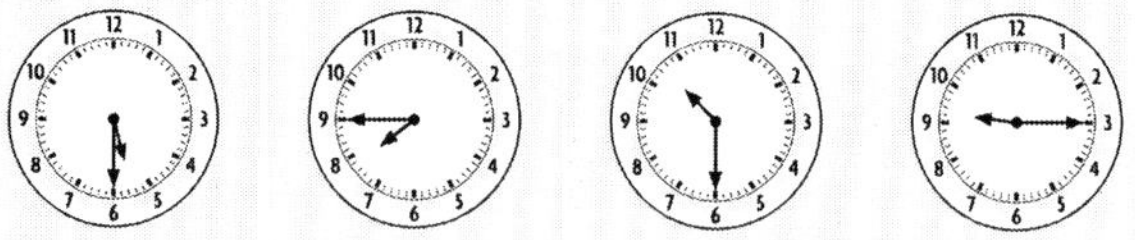

5:30 | 7:45 | 10:30 | 9:15

Draw the hands on the clocks to match the times.

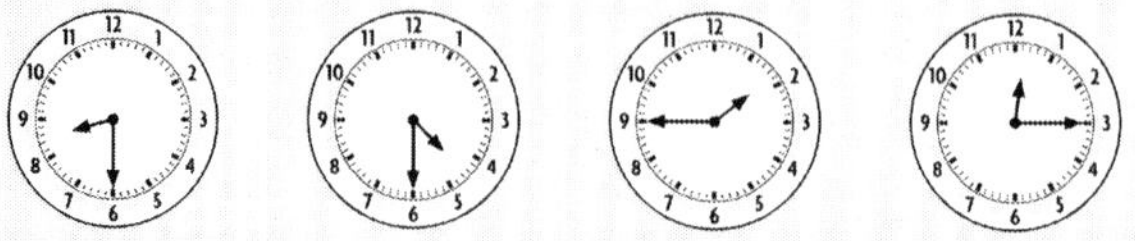

8:30 | 4:30 | 1:45 | 12:15

PAGE 41

Times Ending in :00, :15, :30, :45

Write the time in the blanks below the clocks.

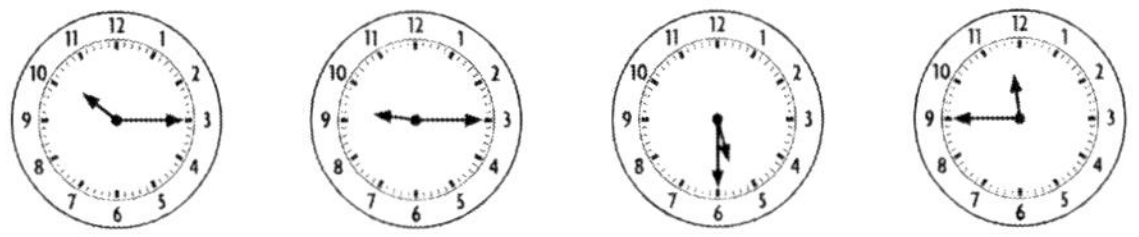

10:15 | 9:15 | 5:30 | 11:45

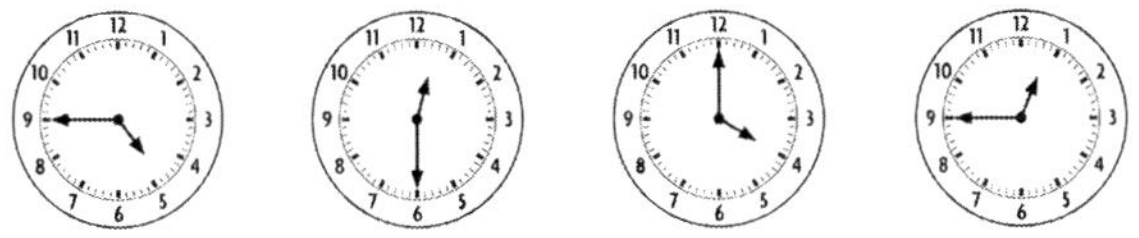

4:45 | 12:30 | 4:00 | 12:45

Draw the hands on the clocks to match the times.

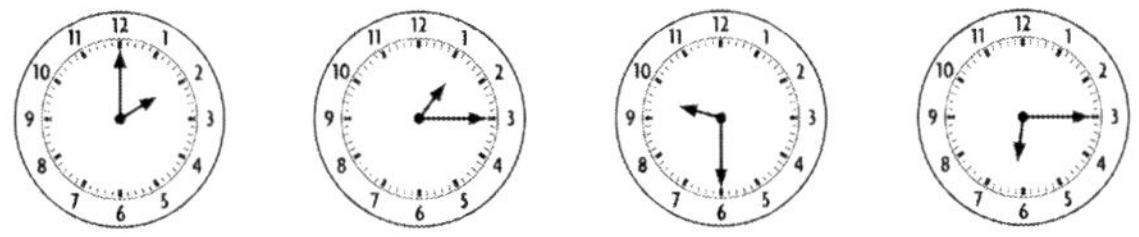

2:00 | 1:15 | 9:30 | 6:15

PAGE 42

Times Ending in :00, :15, :30, :45

Write the time below the clock showing the hands, then write the correct future time below the empty clock and draw the hands.

45 minutes later

6:00 | 6:45

15 minutes later

2:00 | 2:15

Write the time below the clock showing the hands, then write the correct past time below the empty clock and draw the hands.

45 minutes earlier

3:00 | 3:45

PAGE 43

Times Ending in :00, :15, :30, :45

Write the time below the clock showing the hands, then write the correct future time below the empty clock and draw the hands.

30 minutes later

5:00 | 5:30

4 hours later

12:00 | 4:00

Write the time below the clock showing the hands, then write the correct past time below the empty clock and draw the hands.

15 minutes earlier

9:00 | 9:15

PAGE 44

Times Ending in :00, :15, :30, :45

Write the time below the clock showing the hands, then write the correct future time below the empty clock and draw the hands.

15 minutes later

11:00 | 11:15

45 minutes later

1:00 | 1:45

Write the time below the clock showing the hands, then write the correct past time below the empty clock and draw the hands.

30 minutes earlier

10:00 | 10:30

PAGE 45

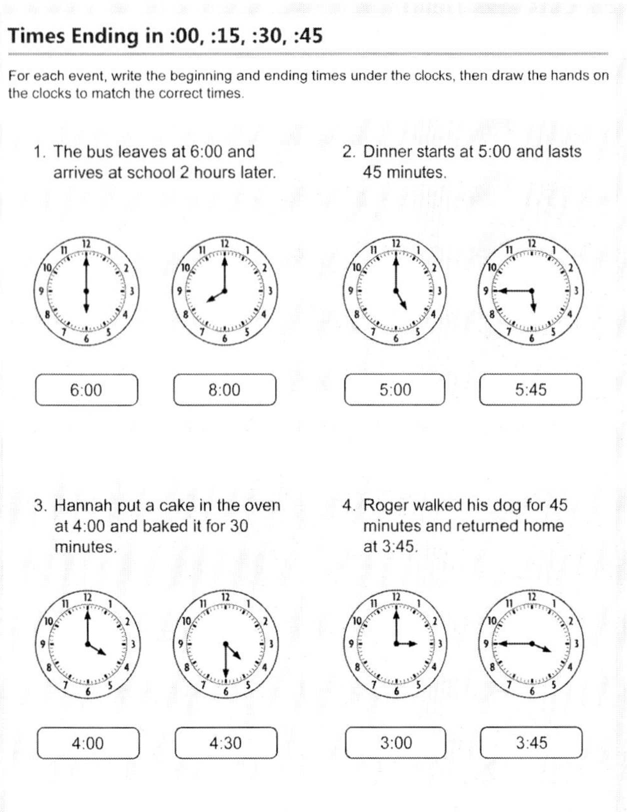

Times Ending in :00, :15, :30, :45

For each event, write the beginning and ending times under the clocks, then draw the hands on the clocks to match the correct times.

1. The bus leaves at 6:00 and arrives at school 2 hours later.

6:00 | 8:00

2. Dinner starts at 5:00 and lasts 45 minutes.

5:00 | 5:45

3. Hannah put a cake in the oven at 4:00 and baked it for 30 minutes.

4:00 | 4:30

4. Roger walked his dog for 45 minutes and returned home at 3:45.

3:00 | 3:45

PAGE 46

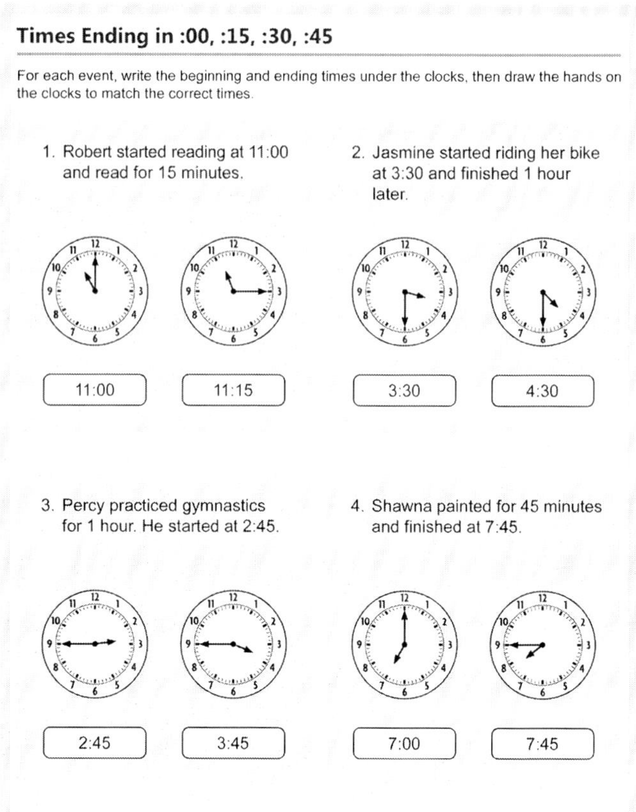

Times Ending in :00, :15, :30, :45

For each event, write the beginning and ending times under the clocks, then draw the hands on the clocks to match the correct times.

1. Robert started reading at 11:00 and read for 15 minutes.

11:00 | 11:15

2. Jasmine started riding her bike at 3:30 and finished 1 hour later.

3:30 | 4:30

3. Percy practiced gymnastics for 1 hour. He started at 2:45.

2:45 | 3:45

4. Shawna painted for 45 minutes and finished at 7:45.

7:00 | 7:45

PAGE 47

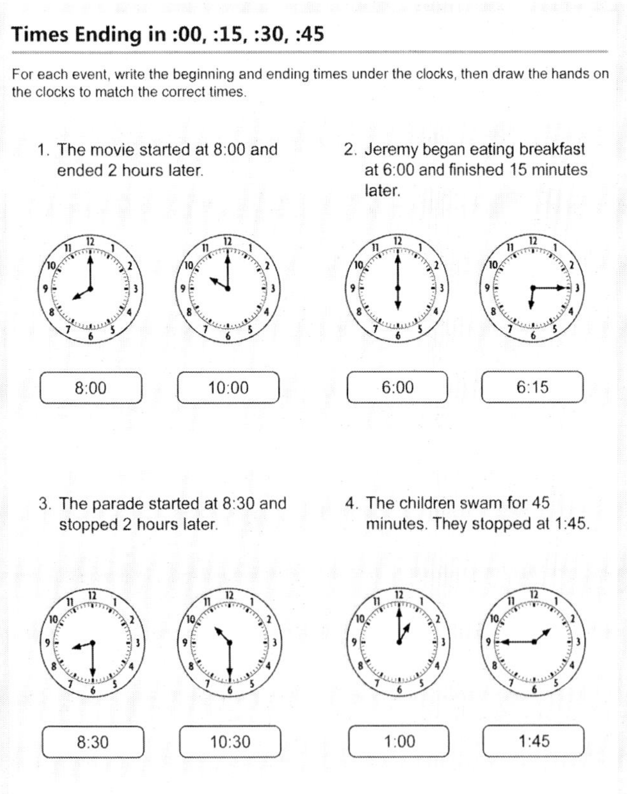

Times Ending in :00, :15, :30, :45

For each event, write the beginning and ending times under the clocks, then draw the hands on the clocks to match the correct times.

1. The movie started at 8:00 and ended 2 hours later.

8:00 | 10:00

2. Jeremy began eating breakfast at 6:00 and finished 15 minutes later.

6:00 | 6:15

3. The parade started at 8:30 and stopped 2 hours later.

8:30 | 10:30

4. The children swam for 45 minutes. They stopped at 1:45.

1:00 | 1:45

PAGE 48

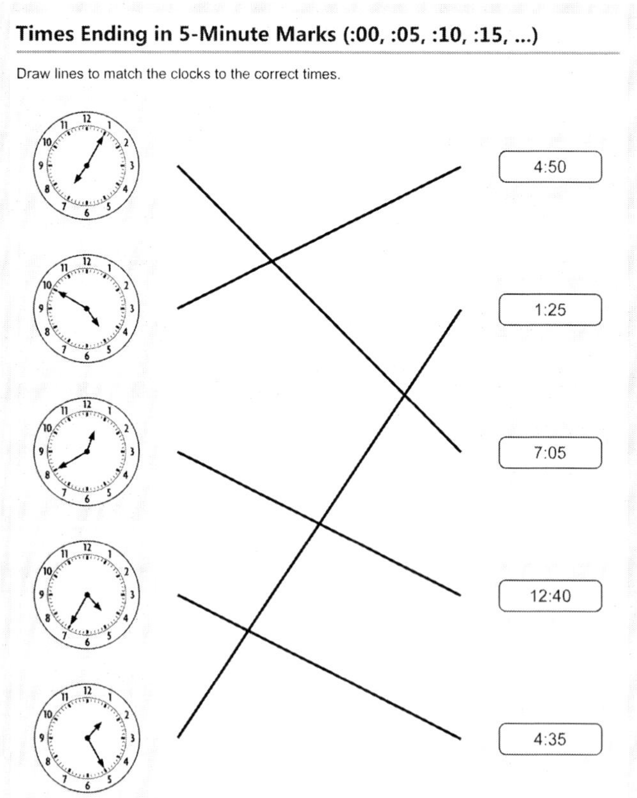

Times Ending in 5-Minute Marks (:00, :05, :10, :15, ...)

Draw lines to match the clocks to the correct times.

PAGE 50

Times Ending in 5-Minute Marks (:00, :05, :10, :15, ...)

Draw lines to match the clocks to the correct times.

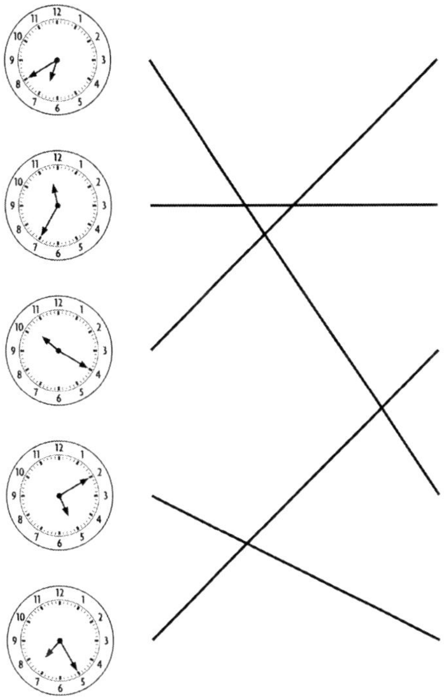

10:20

11:35

7:25

6:40

5:10

PAGE 51

Times Ending in 5-Minute Marks (:00, :05, :10, :15, ...)

Write the time in the blanks below the clocks.

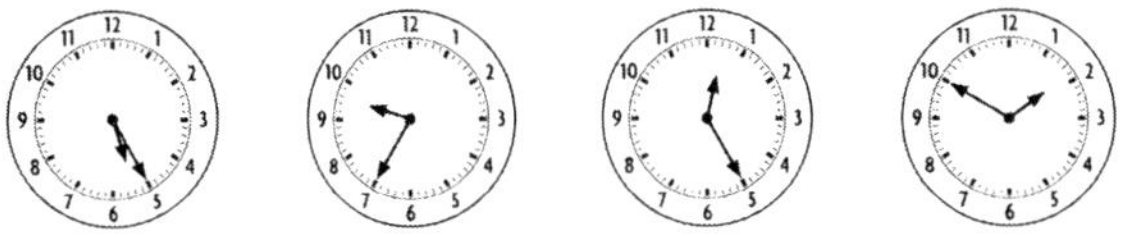

5:25	9:35	12:25	1:50

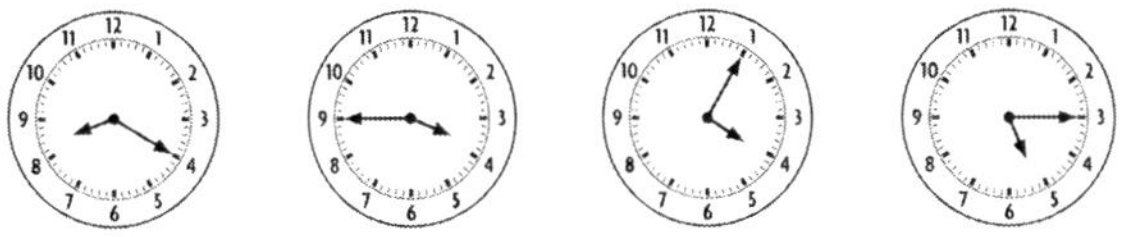

8:20	3:45	4:05	5:15

Draw the hands on the clocks to match the times.

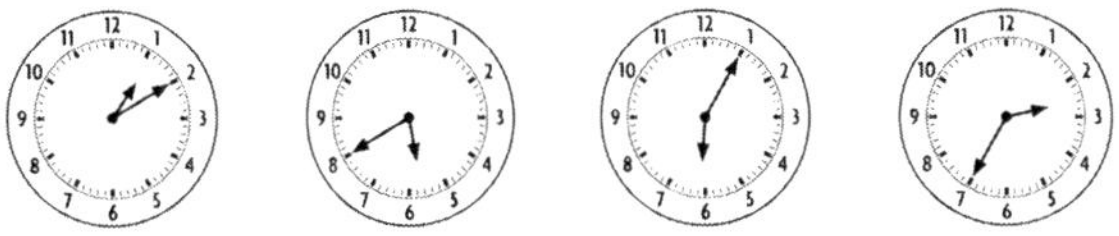

1:10	5:40	6:05	2:35

PAGE 52

Times Ending in 5-Minute Marks (:00, :05, :10, :15, ...)

Write the time in the blanks below the clocks.

1:40	10:10	2:50	4:00

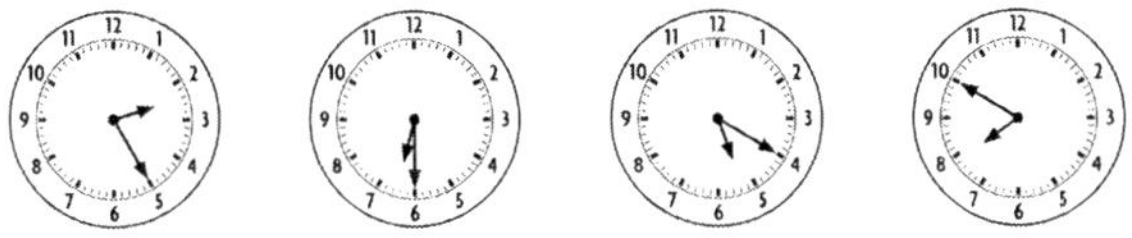

2:25	6:30	5:20	7:50

Draw the hands on the clocks to match the times.

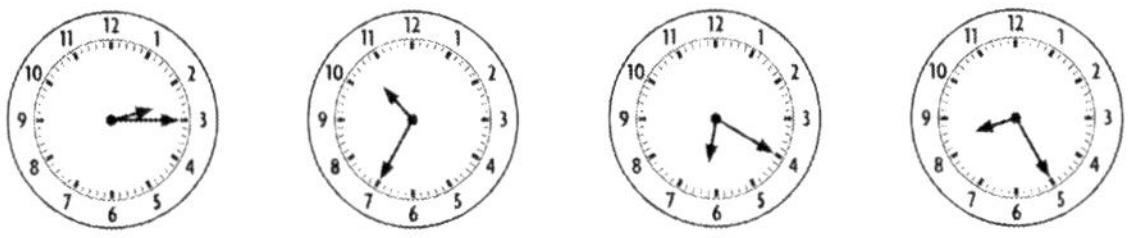

2:15	10:35	6:20	8:25

PAGE 53

Times Ending in 5-Minute Marks (:00, :05, :10, :15, ...)

Write the time in the blanks below the clocks.

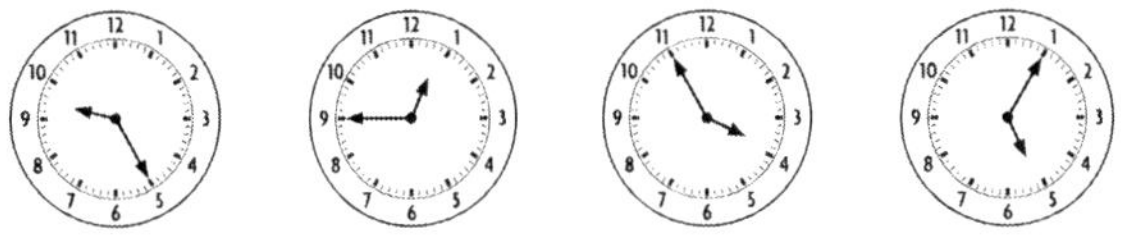

9:25	12:45	3:55	5:05

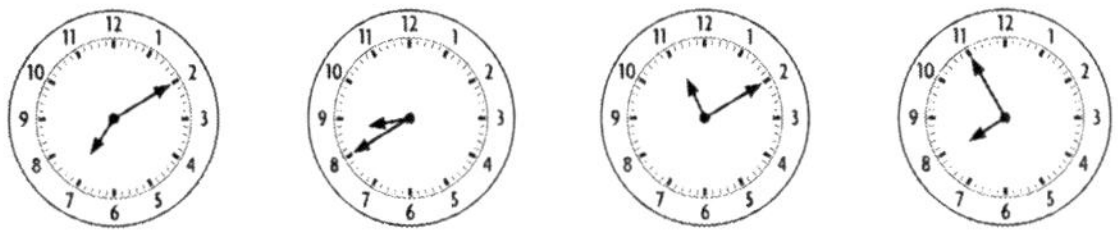

7:10	8:40	11:10	7:55

Draw the hands on the clocks to match the times.

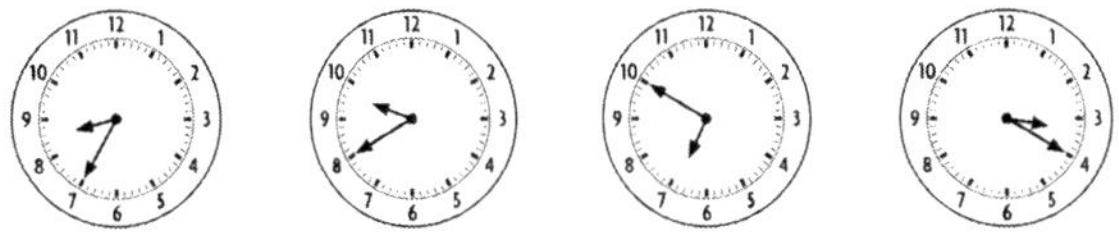

8:35	9:40	6:50	3:20

PAGE 54

Times Ending in 5-Minute Marks (:00, :05, :10, :15, ...)

Write the time in the blanks below the clocks.

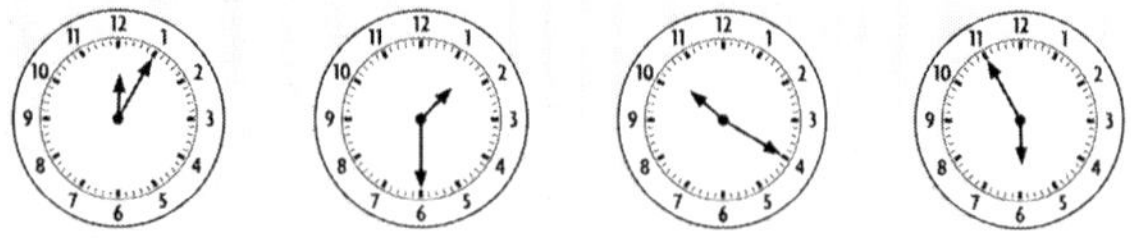

12:05 | 1:30 | 10:20 | 5:55

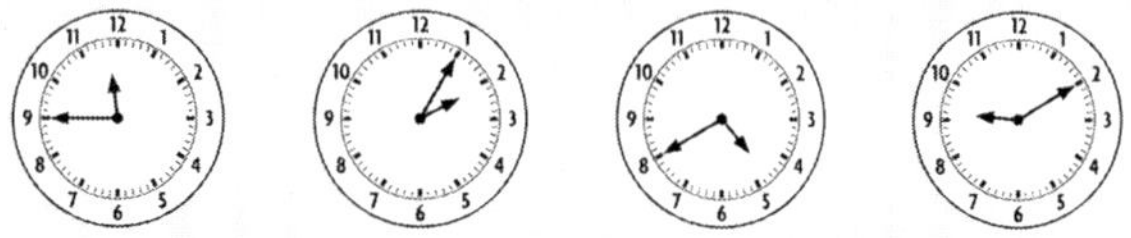

11:45 | 2:05 | 4:40 | 9:10

Draw the hands on the clocks to match the times.

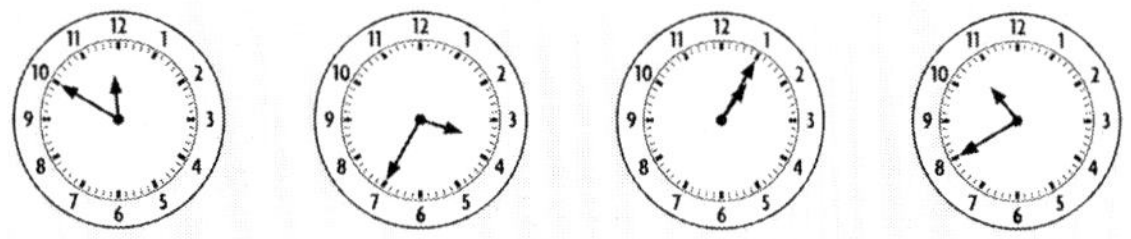

11:50 | 3:35 | 1:05 | 10:40

PAGE 55

Times Ending in 5-Minute Marks (:00, :05, :10, :15, ...)

Write the time in the blanks below the clocks.

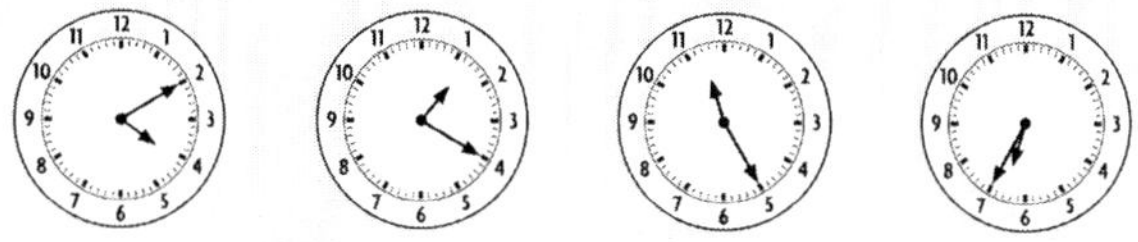

4:10 | 1:20 | 11:25 | 6:35

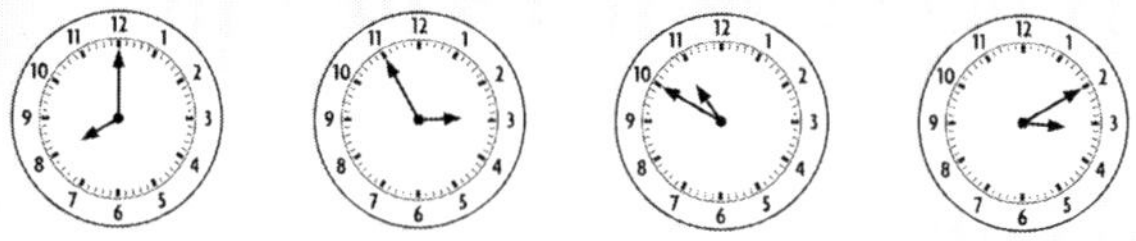

8:00 | 2:55 | 10:50 | 3:10

Draw the hands on the clocks to match the times.

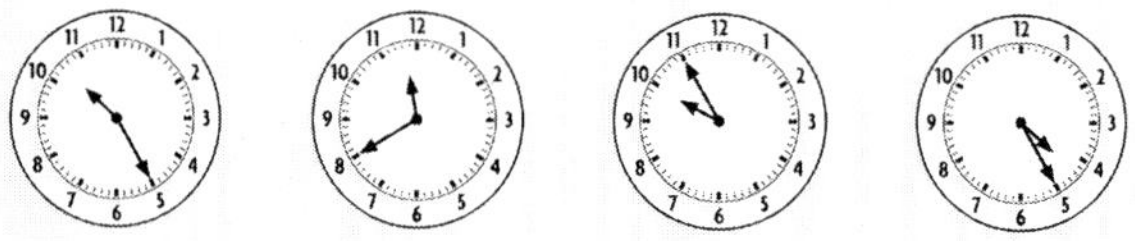

10:25 | 11:40 | 9:55 | 4:25

PAGE 56

Times Ending in 5-Minute Marks (:00, :05, :10, :15, ...)

Write the time in the blanks below the clocks.

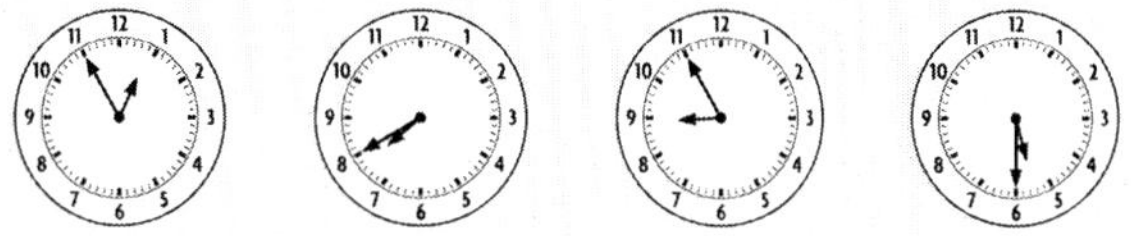

12:55 | 7:40 | 8:55 | 5:30

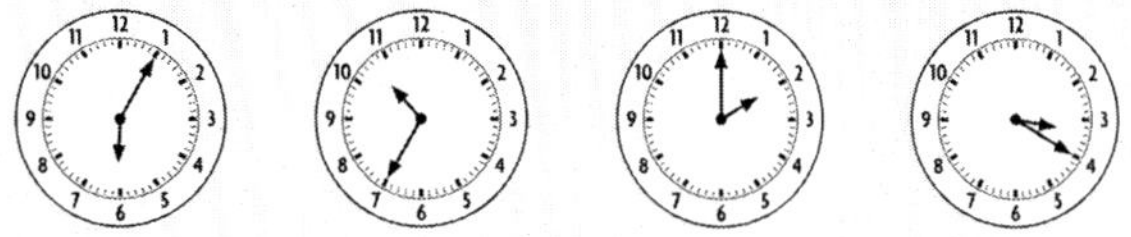

6:05 | 10:35 | 2:00 | 3:20

Draw the hands on the clocks to match the times.

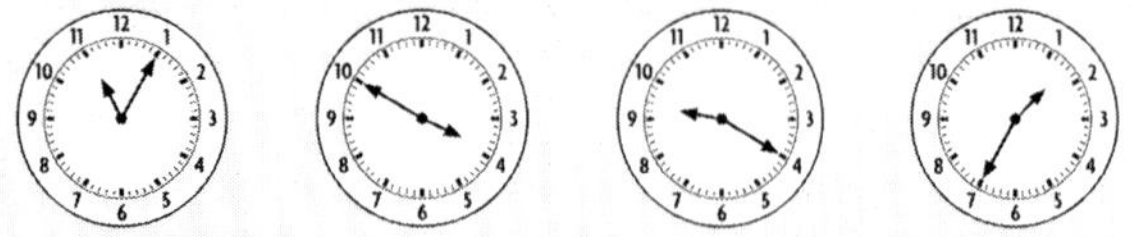

11:05 | 3:50 | 9:20 | 1:35

PAGE 57

Times Ending in 5-Minute Marks (:00, :05, :10, :15, ...)

Write the time in the blanks below the clocks.

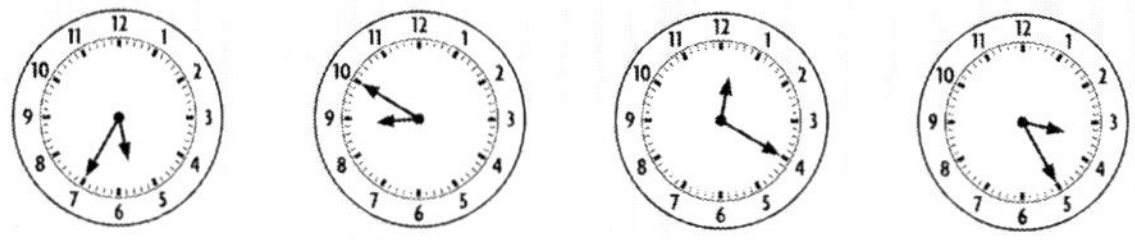

5:35 | 8:50 | 12:20 | 3:25

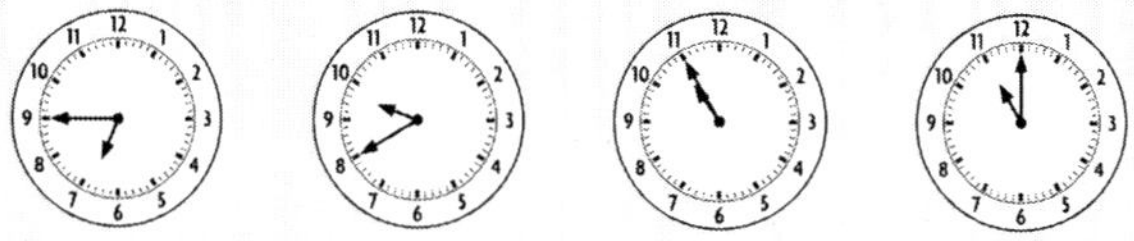

6:45 | 9:40 | 10:55 | 11:00

Draw the hands on the clocks to match the times.

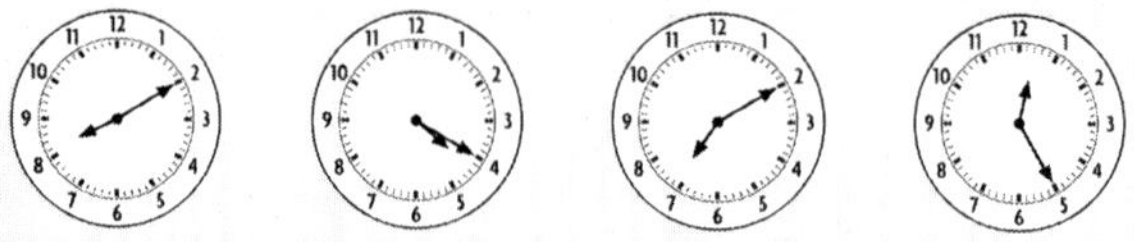

8:10 | 4:20 | 7:10 | 12:25

PAGE 58

Times Ending in 5-Minute Marks (:00, :05, :10, :15, ...)

Write the time in the blanks below the clocks.

11:35 | 5:05 | 9:40 | 4:50

2:15 | 10:35 | 7:25 | 12:40

Draw the hands on the clocks to match the times.

3:55 | 2:25 | 4:05 | 10:10

PAGE 59

Times Ending in 5-Minute Marks (:00, :05, :10, :15, ...)

Write the time below the clock showing the hands, then write the correct future time below the empty clock and draw the hands.

4:05

5 minutes later

4:10

10:30

10 minutes later

10:40

Write the time below the clock showing the hands, then write the correct past time below the empty clock and draw the hands.

8:20

5 minutes earlier

8:25

PAGE 60

Times Ending in 5-Minute Marks (:00, :05, :10, :15, ...)

Write the time below the clock showing the hands, then write the correct future time below the empty clock and draw the hands.

11:00

35 minutes later

11:35

5:50

5 minutes later

5:55

Write the time below the clock showing the hands, then write the correct past time below the empty clock and draw the hands.

6:10

10 minutes earlier

6:20

PAGE 61

Times Ending in 5-Minute Marks (:00, :05, :10, :15, ...)

Write the time below the clock showing the hands, then write the correct future time below the empty clock and draw the hands.

1:30

5 minutes later

1:35

3:05

20 minutes later

3:25

Write the time below the clock showing the hands, then write the correct past time below the empty clock and draw the hands.

7:10

10 minutes earlier

7:20

PAGE 62

PAGE 63

Times Ending in 5-Minute Marks (:00, :05, :10, :15, ...)

For each event, write the beginning and ending times under the clocks, then draw the hands on the clocks to match the correct times.

1. Misha began eating a sandwich at 10:20 and finished eating in 5 minutes.

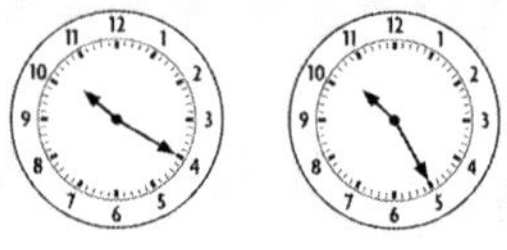

10:20 | 10:25

2. Jeremy practiced batting at 2:30 and stopped 10 minutes later.

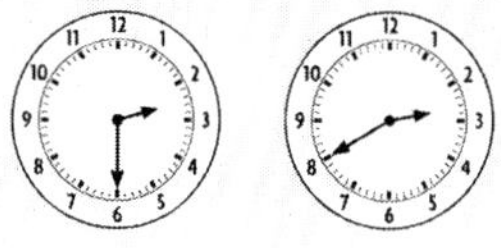

2:30 | 2:40

3. Dell spent 1 hour building a LEGO fort. He started at 4:35.

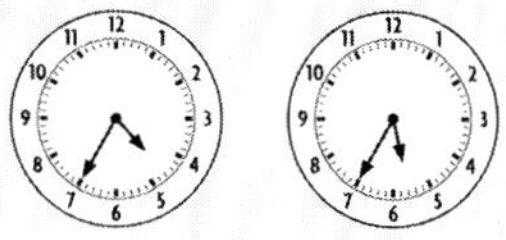

4:35 | 5:35

4. Jane practiced jumping for 5 minutes and stopped at 12:25.

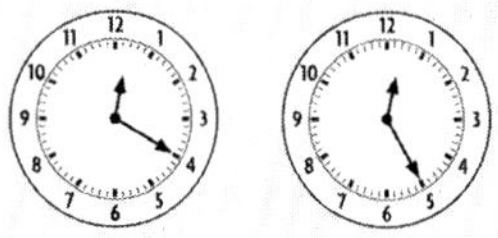

12:20 | 12:25

PAGE 64

Times Ending in 5-Minute Marks (:00, :05, :10, :15, ...)

For each event, write the beginning and ending times under the clocks, then draw the hands on the clocks to match the correct times.

1. The concert began at 6:40 and was over in 2 hours.

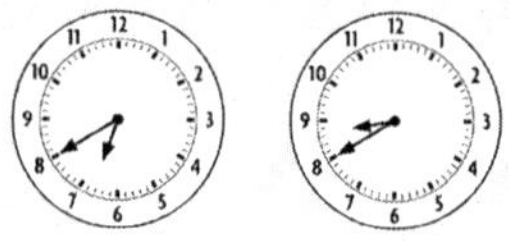

6:40 | 8:40

2. Grista started her worksheet at 5:45 and finished in 5 minutes.

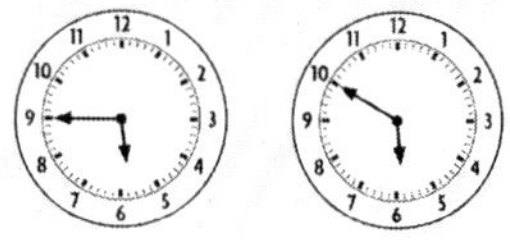

5:45 | 5:50

3. Abby ran for 10 minutes. She started at 9:20.

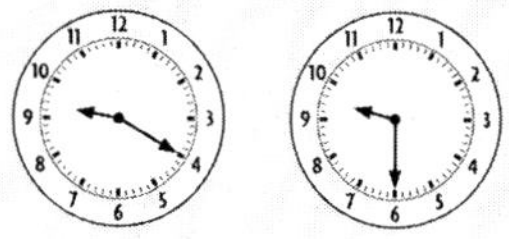

9:20 | 9:30

4. The fan stopped spinning at 7:55. It was on for 5 minutes.

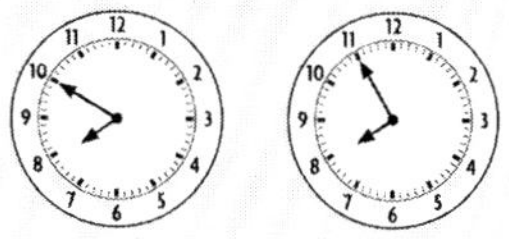

7:50 | 7:55

PAGE 65

Times Ending in 5-Minute Marks (:00, :05, :10, :15, ...)

For each event, write the beginning and ending times under the clocks, then draw the hands on the clocks to match the correct times.

1. Hugo left his house 1:20 and walked for 5 minutes to his friend's house.

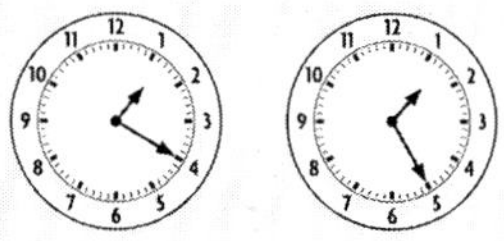

1:20 | 1:25

2. Shannon started cleaning her room at 4:00 and finished 35 minutes later.

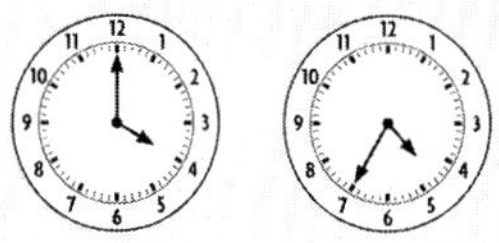

4:00 | 4:35

3. Rita played video games with her little brother at 11:50 and they played for 1 hour.

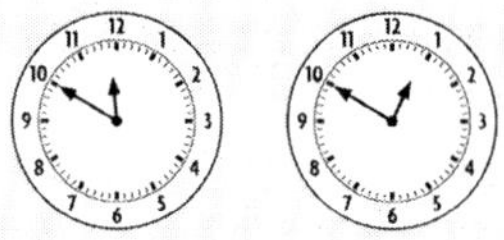

11:50 | 12:50

4. Jillian's cookies were finished baking at 5:30. They were in the oven for 10 minutes.

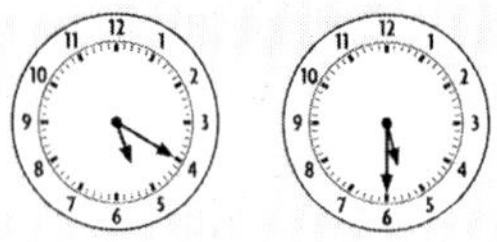

5:20 | 5:30

PAGE 66

Times Ending in 5-Minute Marks (:00, :05, :10, :15, ...)

For each event, write the beginning and ending times under the clocks, then draw the hands on the clocks to match the correct times.

1. Isabelle rode her horse for 1 hour. She started at 4:40.

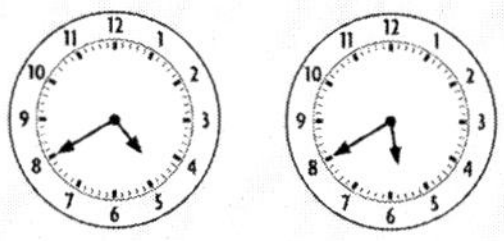

4:40 | 5:40

2. Mark started eating an apple at 11:20 and finished in 5 minutes.

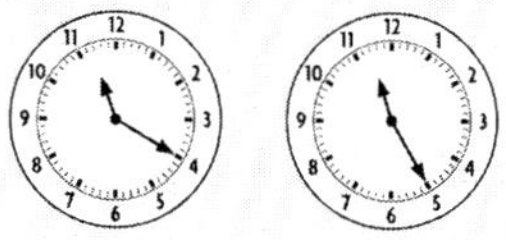

11:20 | 11:25

3. Erin brushed her hair for 10 minutes. She started at 8:10.

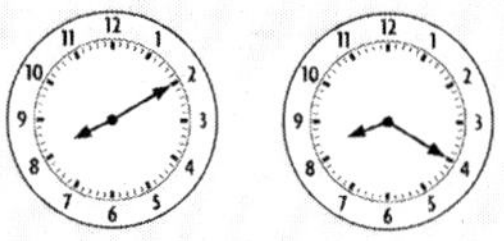

8:10 | 8:20

4. The popcorn finished popping at 6:15. It popped for 5 minutes.

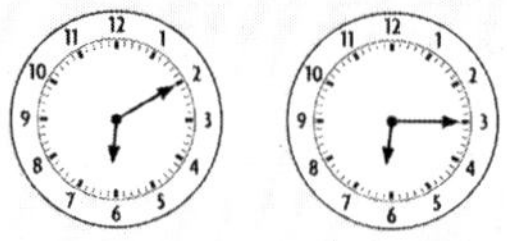

6:10 | 6:15

Times Ending in 1-Minute Marks (:00, :01, :02, :03, ...)

Draw lines to match the clocks to the correct times.

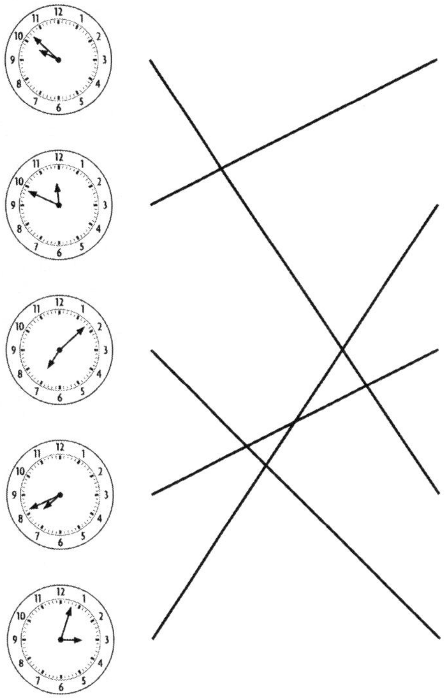

11:49

3:03

7:41

9:52

7:08

PAGE 68

Times Ending in 1-Minute Marks (:00, :01, :02, :03, ...)

Draw lines to match the clocks to the correct times.

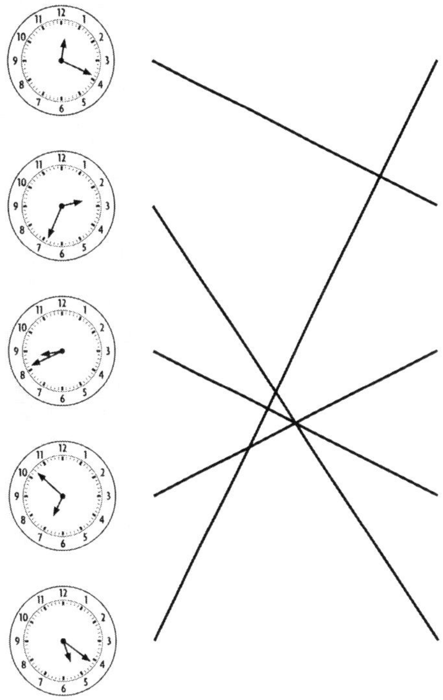

5:21

12:19

6:52

8:41

2:34

PAGE 69

Times Ending in 1-Minute Marks (:00, :01, :02, :03, ...)

Write the time in the blanks below the clocks.

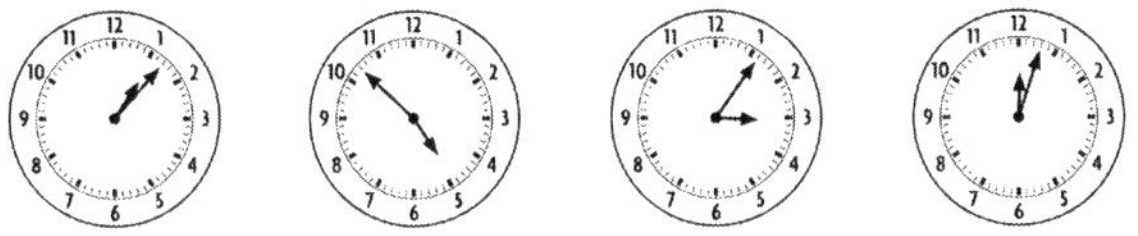

1:07 4:52 3:06 12:03

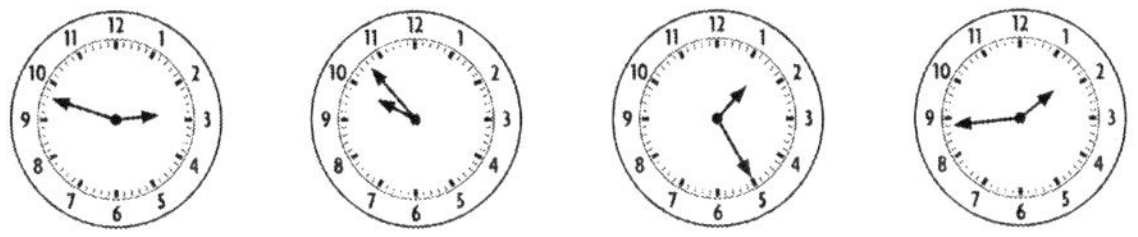

2:48 9:53 1:25 1:44

Draw the hands on the clocks to match the times.

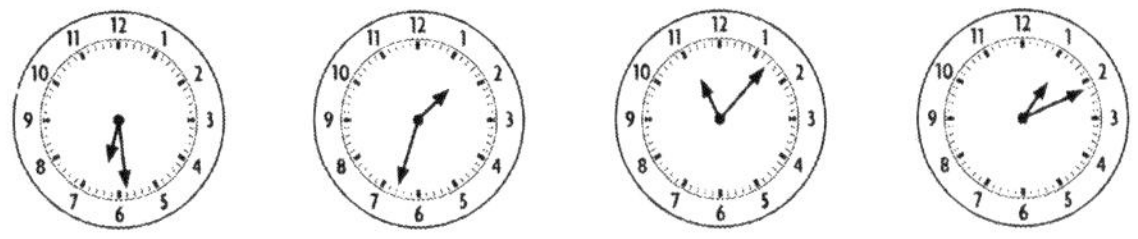

6:29 1:33 11:07 1:11

PAGE 70

Times Ending in 1-Minute Marks (:00, :01, :02, :03, ...)

Write the time in the blanks below the clocks.

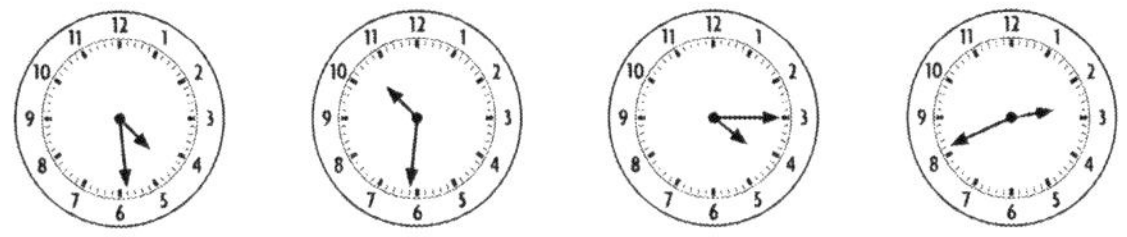

4:29 10:31 4:15 2:41

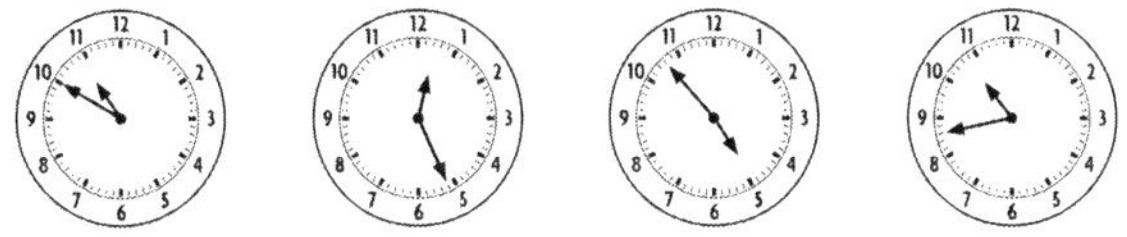

10:50 12:26 4:53 10:43

Draw the hands on the clocks to match the times.

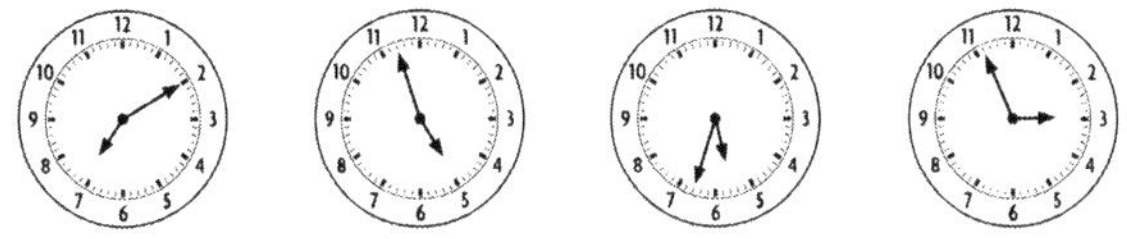

7:10 4:57 5:33 2:56

PAGE 71

Times Ending in 1-Minute Marks (:00, :01, :02, :03, ...)

Write the time in the blanks below the clocks.

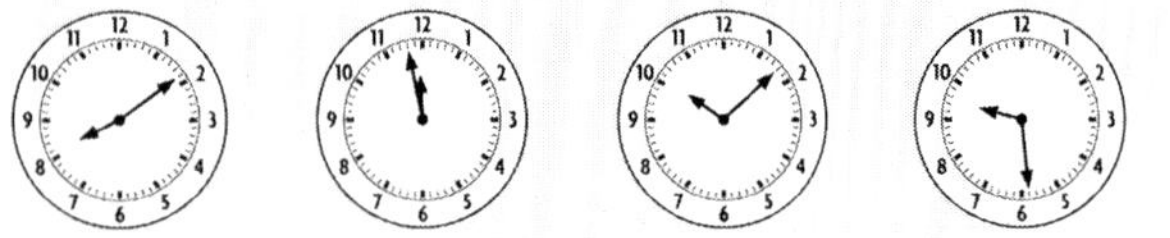

8:09 | 11:58 | 10:08 | 9:29

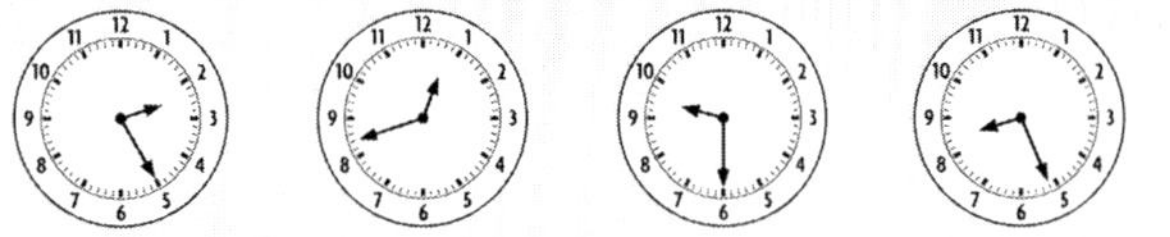

2:25 | 12:42 | 9:30 | 8:26

Draw the hands on the clocks to match the times.

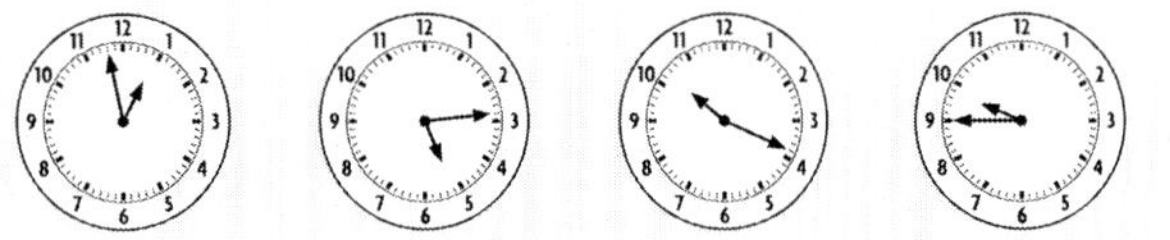

12:58 | 5:14 | 10:19 | 9:45

PAGE 72

Times Ending in 1-Minute Marks (:00, :01, :02, :03, ...)

Write the time in the blanks below the clocks.

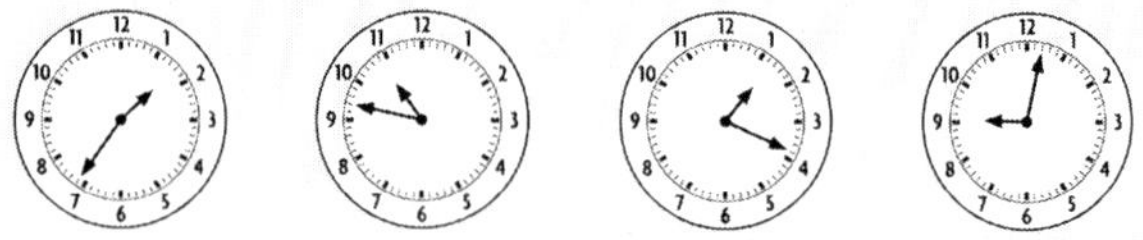

1:36 | 10:47 | 1:19 | 9:02

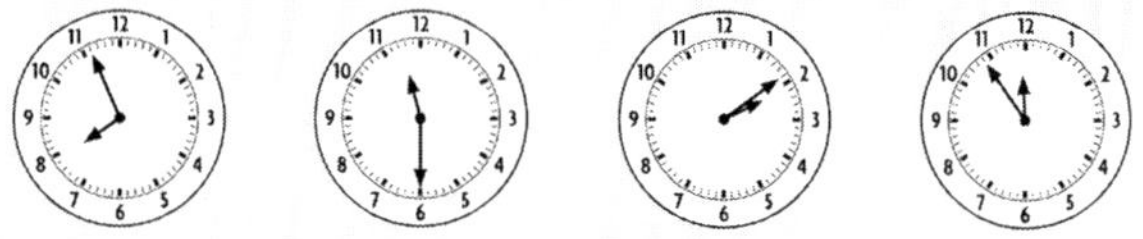

7:56 | 11:30 | 2:09 | 11:54

Draw the hands on the clocks to match the times.

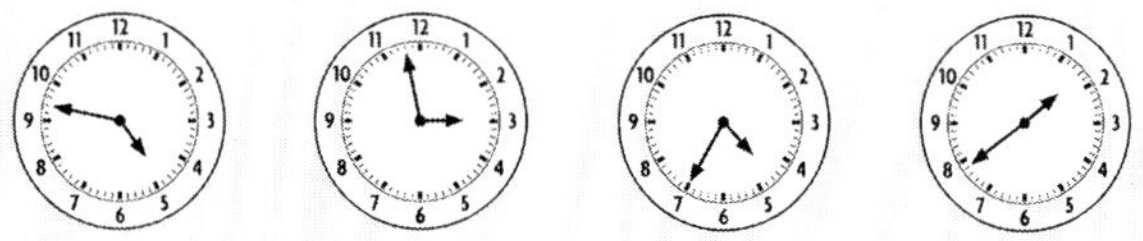

4:47 | 2:58 | 4:35 | 1:39

PAGE 73

Times Ending in 1-Minute Marks (:00, :01, :02, :03, ...)

Write the time in the blanks below the clocks.

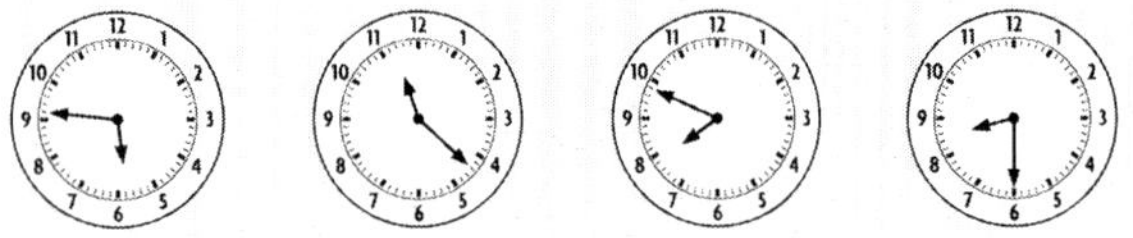

5:46 | 11:22 | 7:49 | 8:30

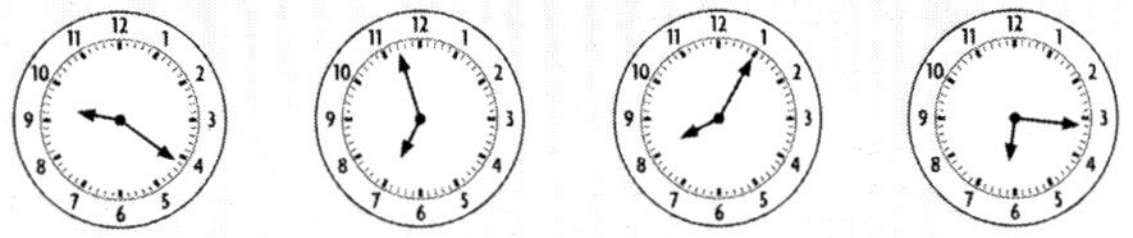

9:21 | 6:57 | 8:05 | 6:16

Draw the hands on the clocks to match the times.

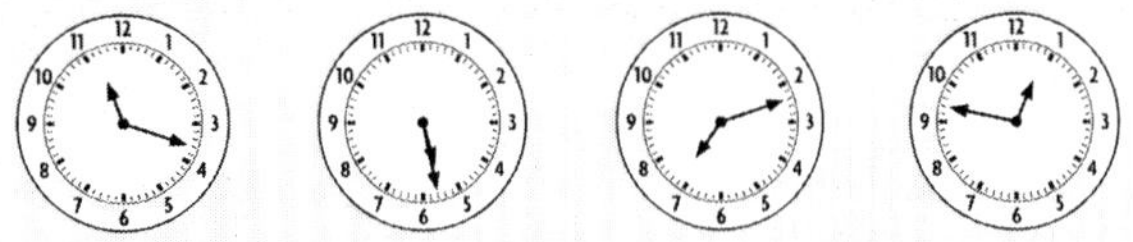

11:18 | 5:28 | 7:12 | 12:47

PAGE 74

Times Ending in 1-Minute Marks (:00, :01, :02, :03, ...)

Write the time in the blanks below the clocks.

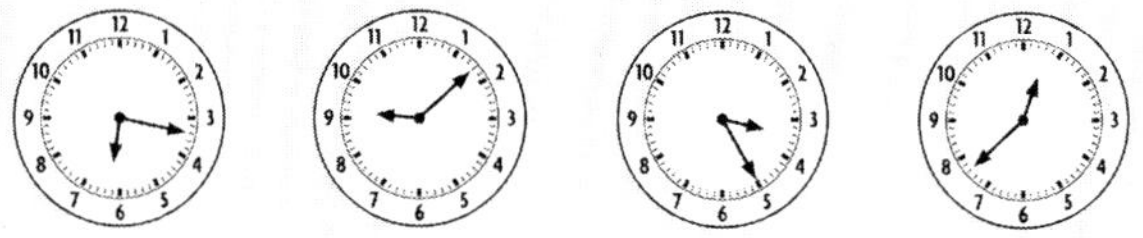

6:17 | 9:08 | 3:25 | 12:38

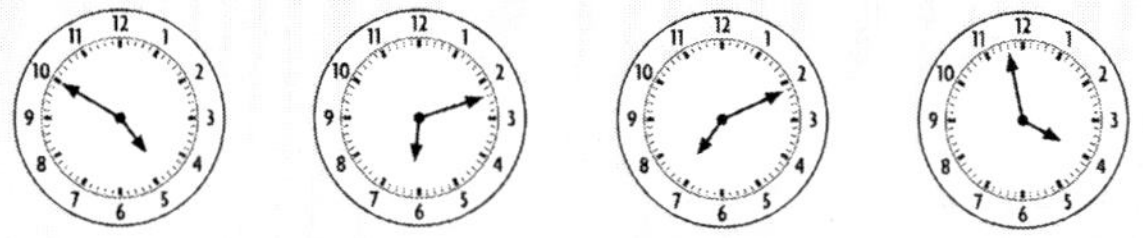

4:50 | 6:12 | 7:11 | 3:58

Draw the hands on the clocks to match the times.

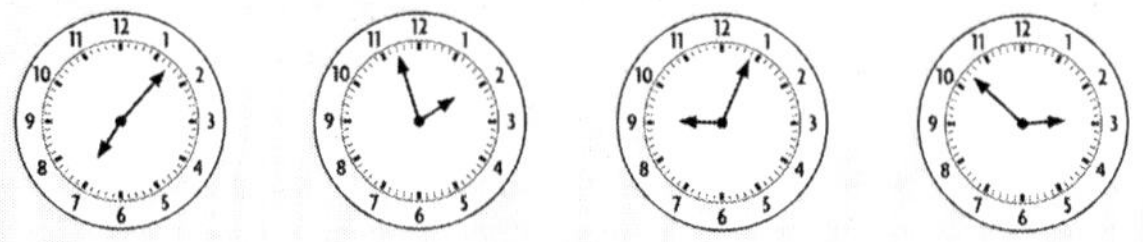

7:07 | 1:57 | 9:04 | 2:52

PAGE 75

Times Ending in 1-Minute Marks (:00, :01, :02, :03, ...)

Write the time in the blanks below the clocks.

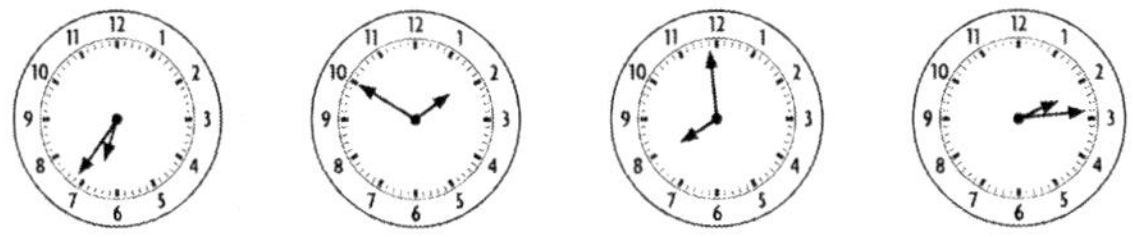

6:36 | 1:50 | 7:59 | 2:14

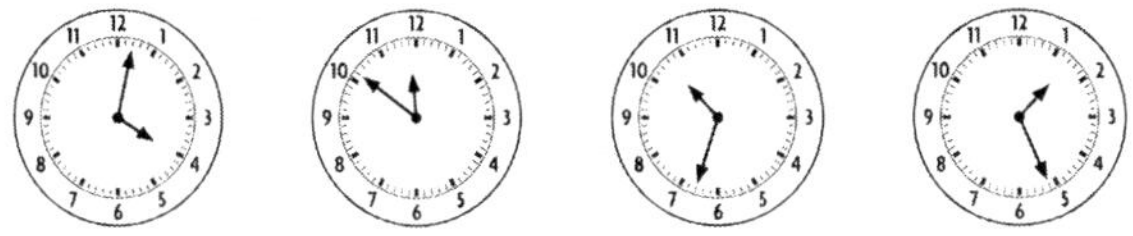

4:02 | 11:51 | 10:33 | 1:26

Draw the hands on the clocks to match the times.

6:49 | 2:23 | 5:18 | 8:22

PAGE 76

Times Ending in 1-Minute Marks (:00, :01, :02, :03, ...)

Write the time in the blanks below the clocks.

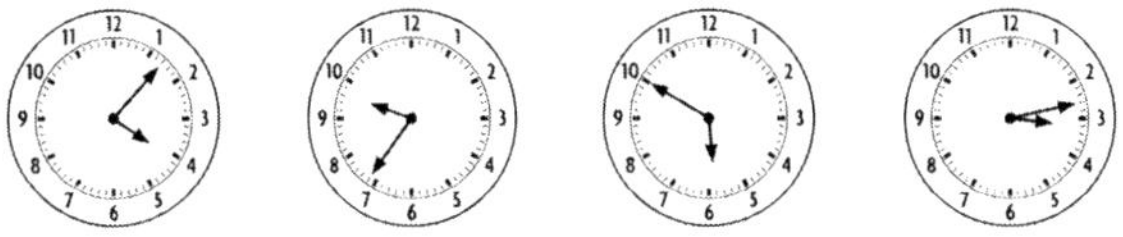

4:07 | 9:36 | 5:50 | 3:13

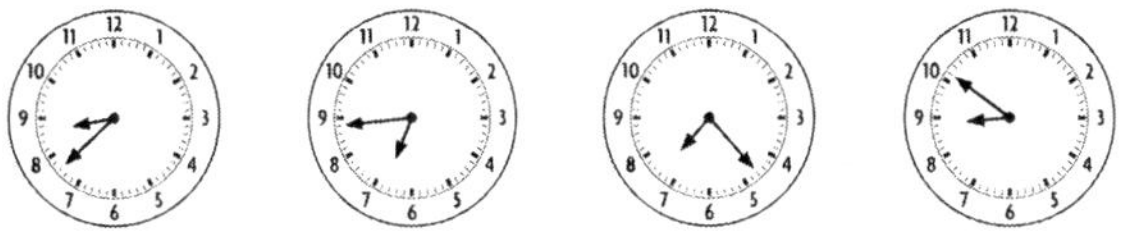

8:38 | 6:44 | 7:23 | 8:51

Draw the hands on the clocks to match the times.

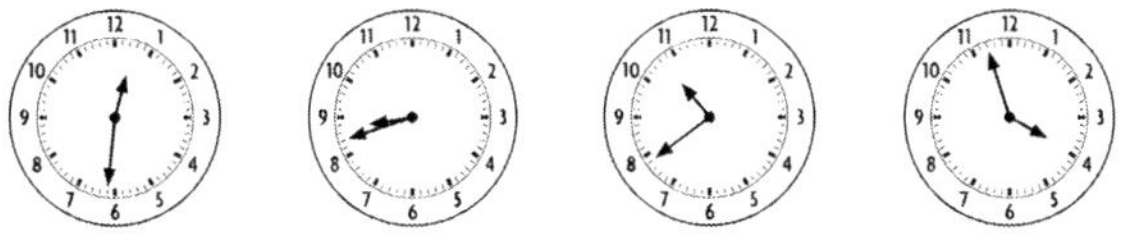

12:31 | 8:42 | 10:39 | 3:57

PAGE 77

Times Ending in 1-Minute Marks (:00, :01, :02, :03, ...)

Write the time below the clock showing the hands, then write the correct future time below the empty clock and draw the hands.

1 minute later

12:30 | 12:31

3 minutes later

5:10 | 5:13

Write the time below the clock showing the hands, then write the correct past time below the empty clock and draw the hands.

1 minute earlier

1:14 | 1:15

PAGE 78

Times Ending in 1-Minute Marks (:00, :01, :02, :03, ...)

Write the time below the clock showing the hands, then write the correct future time below the empty clock and draw the hands.

4 minutes later

3:20 | 3:24

12 minutes later

9:00 | 9:12

Write the time below the clock showing the hands, then write the correct past time below the empty clock and draw the hands.

3 minutes earlier

4:10 | 4:13

PAGE 79

Times Ending in 1-Minute Marks (:00, :01, :02, :03, ...)

Write the time below the clock showing the hands, then write the correct future time below the empty clock and draw the hands.

9:34

1 hour later

10:34

7:00

8 minutes later

7:08

Write the time below the clock showing the hands, then write the correct past time below the empty clock and draw the hands.

3:42

1 hour earlier

4:42

PAGE 80

Times Ending in 1-Minute Marks (:00, :01, :02, :03, ...)

For each event, write the beginning and ending times under the clocks, then draw the hands on the clocks to match the correct times.

1. The show started at 2:00 and ended 23 minutes later.

2:00 | 2:23

2. The friends talked on the phone for 31 minutes. They started at 5:00.

5:00 | 5:31

3. Jacob finished the math worksheet in 4 minutes. He started at 9:20.

9:20 | 9:24

4. Jessica drank her orange juice in 3 minutes. She was finished at 7:05.

7:02 | 7:05

PAGE 81

Times Ending in 1-Minute Marks (:00, :01, :02, :03, ...)

For each event, write the beginning and ending times under the clocks, then draw the hands on the clocks to match the correct times.

1. James cleaned the table in 3 minutes. He started at 10:05.

10:05 | 10:08

2. Rose began singing at 1:10 and finished the song in 4 minutes.

1:10 | 1:14

3. Luke started dancing at 5:02. He danced for 10 minutes.

5:02 | 5:12

4. Mara ate a cookie in 1 minute. She was finished at 12:24.

12:23 | 12:24

PAGE 82

Times Ending in 1-Minute Marks (:00, :01, :02, :03, ...)

For each event, write the beginning and ending times under the clocks, then draw the hands on the clocks to match the correct times.

1. The ant began to walk across the sidewalk at 8:22 and arrived at the other side 1 minute later.

8:22 | 8:23

2. The cat started chasing a mouse at 3:20 and gave up 4 minutes later.

3:20 | 3:24

3. The chef kneaded dough for 8 minutes. He began kneading at 2:20.

2:20 | 2:28

4. The students did jumping jacks for 3 minutes. They stopped at 4:05.

4:02 | 4:05

PAGE 83

Times Ending in 1-Minute Marks (:00, :01, :02, :03, ...)

For each event, write the beginning and ending times under the clocks, then draw the hands on the clocks to match the correct times.

1. The dog began to chew on a bone at 12:00 and stopped 7 minutes later.

12:00 | 12:07

2. The students lined up to walk outside at 10:30. They were on the playground 4 minutes later.

10:30 | 10:34

3. The children spent 2 minutes washing their hands. They started at 2:30.

2:30 | 2:32

4. Daniel took 3 minutes to water the plant and stopped at 1:13.

1:10 | 1:13

PAGE 84

If you found this book helpful, we'd really appreciate your feedback.
Please consider leaving a review!

Made in United States
North Haven, CT
05 November 2026

21772477R00061